The
People's
Spaceship

The
People's
Spaceship

NASA, the Shuttle Era, and Public Engagement after Apollo

Amy Paige Kaminski

University of Pittsburgh Press

Published by the University of Pittsburgh Press, Pittsburgh, Pa., 15260
Copyright © 2024, University of Pittsburgh Press
Manufactured in the United States of America
Printed on acid-free paper
10 9 8 7 6 5 4 3 2 1

Cataloging-in-Publication data is available from the Library of Congress

ISBN 13: 978-0-8229-4766-0
ISBN 10: 0-8229-4766-8

Cover art: Based on a NASA mission insignia patch for STS-27.

Cover design: Alex Wolfe

For Mom and Dad and Maya

Contents

Preface

The People's Spaceship: NASA, the Shuttle Era, and Public Engagement after Apollo signifies for me the melding of personal dreams past and present. I grew up with the goal of becoming a space shuttle astronaut. Shuttle posters adorned my childhood bedroom, and I loved learning about the mission crews and how the vehicle operated. During my undergraduate years, however, I realized that my genuine passion entailed not exploring space per se but considering the social, political, and cultural imperatives and questions surrounding this bold human endeavor. Why have some nations chosen to invest in sending people or robotic emissaries into space? What do space initiatives like the shuttle achieve for these countries and their citizens? How can spacefaring nations make the benefits of space activities equitable and broadly accessible? Who will get to participate in space flight in the future and how, especially as private companies and individuals play a greater role in using and exploring space?

Although I abandoned my aspiration to fly to the stars, my fascination with the shuttle and my commitment to contributing to space exploration never ebbed. I have spent more than two decades grappling with issues like these as a space policy and public engagement professional, most recently at NASA, and as a scholar. As the shuttle program came to an end, I attended the "retirement ceremony" of orbiters *Discovery* and *Enterprise* at the Smithsonian National Air and Space Museum's Steven F. Udvar-Hazy Center in Chantilly, Virginia, in 2012. There, I could not help but reflect on the vehicle's legacy and the impression it had made on me and surely countless others. I also thought back to watching IMAX movies filmed by shuttle astronauts, collecting mission patches, and being inspired by my middle school teacher, who had been a finalist for the Teacher in Space Project. It was evident to me that NASA had put a great deal of thought into how it connected the shuttle with the people of the world, and especially citizens of the United States. I wondered about how the space agency had made the choices it did about engaging with the American public in the shuttle era. What did they—we—mean to NASA and the shuttle program, and why?

This book is the answer to that curiosity. It is the result of extensive historical research. Written records that informed this project include internal NASA memoranda capturing discussions of issues and decisions related to engaging various publics with the shuttle, NASA fact sheets and brochures about the shuttle, speeches made by NASA officials to external groups, and correspondence between NASA officials and individual citizens concerning the shuttle program. Most of the primary documents I examined come from the NASA Historical Reference Collection at NASA Headquarters in Washington, DC. Others reside in the Johnson Space Center History Collection, stored at the University of Houston–Clear Lake Archives in Texas. NASA's eight-volume *Exploring the Unknown: Selected Documents in the History of the US Civil Space Program* contains reprints of key historical documents that also proved useful in my research.[1]

Presidential policy statements as well as reports by NASA advisory committees and the *Challenger* and *Columbia* accident investigation commissions shed light on some of the external directives and pressures that drove NASA's decisions about public engagement and how its choices changed over time. I reviewed records of congressional hearings capturing debates about the post-Apollo human space flight program and the origins of NASA's mandate to disseminate information about its activities. Examining public opinion polls, articles in national and local news publications, and magazine and newspaper advertisements with shuttle-related themes helped me understand public views of the shuttle program that both resulted from and drove NASA's public engagement efforts.

Original interviews with more than two dozen people connected to the shuttle program enhanced this project by providing background, views, and anecdotes that documents did not or could not provide. Interviewees included shuttle program officials, astronauts, NASA administrators, NASA public affairs and outreach officers, journalists, my middle school teacher, and others who worked with NASA to advance the agency's commitment to public interaction. Talking directly with each of them provided candid perspectives on issues the agency faced in its efforts to engage different groups with the shuttle that would not necessarily have been captured in official agency documentation. I also reviewed transcripts of interviews with NASA officials conducted by other researchers. These transcripts are accessible through the Kennedy Space Center Oral History Project and the Johnson Space Center Oral History Project.

Lastly, I wish to note that the insights I offer in this book are undoubtedly also influenced by the tacit understanding of NASA's culture and perspectives that I gained through the years I have spent shaping NASA policies, programs, and public engagement projects. These experiences, combined with my academic training in science and technology studies—a field that views these enterprises as subject to human choice, just like any other—have positioned me to unearth and tell a multifaceted story of why and how NASA strived for four decades to share the shuttle, quite literally, with a wide range of organizations, groups, and individuals within the United States and around the world. Except where noted, all opinions and interpretations are my own and do not necessarily reflect the views of NASA.

Acknowledgments

Numerous caring and knowledgeable people eased my path to produce this work and are the foundation from which it came to fruition. I am indebted and forever grateful to all who supported me, whether by helping me work through intellectual complexities, holding out a light as I navigated the depths of various archives and research materials, providing gifts of time or financial resources, celebrating with me the achievement of milestones along the way, or commiserating when the going got tough. My most heartfelt thanks go to my family. My daughter, Maya, was six years old when I began this project; she is now a young adult and has been my greatest motivation to complete the journey. I also cannot express enough my appreciation for the support I have received from my parents, James and Cheryl Snyder, and my sisters, Stacey and Randi. My grandparents, none of whom are still living, provided a constant source of inspiration for me to reach my goal.

Many people provided guidance and served as sounding boards. I extend infinite thanks to Sonja Schmid at Virginia Tech, who committed untold numbers of hours for discussions and reviews concerning the project that became this book. Barbara Allen, Gary Downey, and Richard Hirsh provided helpful comments on drafts of this work. Historian Roger Launius contributed constructive insights. My longtime mentor and friend Bruce Lewenstein, of Cornell University, indulged me in invaluable discussions about my research. Daniel Breslau, Michael Dennis, and David Onkst generously provided feedback. Saul Halfon, Matthew Hersch, David Nye, David Tomblin, Janet Vertesi, and Matthew Wisnioski, and the "Albatrosses" of the Society for the History of Technology also helped me formulate ideas early on. My former classmates at Virginia Tech, including Kelley Boyer, Claire Cuccio, the late Phil Egert, Mel Eulau, Jen Henderson, Lee Ann Mawler, Stephanie Mawler, Sterling Mullis, and David Winyard, were wonderful sources of ideas and encouragement. I will not soon forget the lunch dates during which Linda Billings, Ellen McCallie, and Shali Mohleji took the time to talk through my research ideas with me.

Still others made it possible for me to conduct my research. At NASA Headquarters in Washington, DC, Bill Barry, Colin Fries, John Hargenrader, the late Jane Odom, and Liz Suckow facilitated access to the NASA Historical Reference Collection's rich trove of primary source documents. At NASA Johnson Space Center's history office at the University of Houston–Clear Lake, archivists Regina (Jean) Grant and Lauren Meyers graciously supported my two-day, whirlwind visit to the facility. The NASA Headquarters Library team, including the late Craig Levin, Lee Shapiro, and Rick Spencer, provided reading suggestions and good cheer on my trips to collect background materials. At Virginia Tech, Debbie Cash and Bruce Pencek aided me in navigating the university's library system and databases to find valuable source information.

I also extend a great deal of thanks to those I interviewed for sharing their personal experiences with NASA. It was especially poignant to speak with Pennsylvania Teacher-in-Space finalist Pat Palazzolo, who was my own teacher several decades ago. I owe a debt of gratitude to Doug Peterson, who organized and took me on a tour—my first—of Johnson Space Center. Maureen Muncy of NASA's legislative affairs office helped me figure out where to look to unearth the history of the National Aeronautics and Space Act's "dissemination of information" clause. John Logsdon provided a draft chapter from his then forthcoming book on President Richard Nixon's role in the space program. Thor Hogan shared data he collected while writing his dissertation on the space program. Jeffrey Philpott supplied a copy of his thesis on the *Challenger* accident. Thanks also go to Heather Anderson, Aggie Kobrin, and Rod Pyle for helping with image permissions; to Michael Mealing, Brian Odom, and Scott Pace for filling in a few details needed to complete the manuscript; and to Rob Rogers and Pedro Szekely for allowing me to use their artwork in this book. I am also grateful to Adam Greenstone, Clevette Lee, and Katie Spear of NASA's Office of the General Counsel, who provided government ethics advice on writing and presenting my research outside of NASA.

My thanks go to additional colleagues who encouraged me to complete this work: Waleed Abdalati, Gale Allen, Marc Allen, Louis Barbier, Shelly Canright, Cassie Conley, Sarah DeWitt, Brian Dewhurst, Walt Engelund, Jens Feeley, the late Mike Freilich (and wife Shoshannah), Victoria Friedensen, Teresa Fryberger, Steve Garber, Mike Green, Jenn Gustetic, Garth Henning, Rebecca Spyke Keiser, Jason Kessler, Jonathan Krezel, Rich Leshner, Alex MacDonald, Kathy Nado, Michael New, Zach Pirtle, Julie Pollitt, Jonathan Rall, Sarah Becky Ramsey, Tara Ruttley, Lea Shanley, Paul Shawcross, Ellen Stofan, Alotta Taylor, and Cynthia Thomas. I am also indebted to many other special people in my life, including Kirsten Armstrong, Holly Degn, Irene Kariampuzha, Yvette Neisser, Valerie Pratt, Michael Sanghun Rhee, the late Frank Sietzen, Michelle Treistman, Richard Wong, and Pablo Zylberglait, who offered support and made me laugh along the way.

Lastly, I offer deep gratitude to Abby Collier, Amy Sherman, Alex Wolfe, and the rest of the talented team at the University of Pittsburgh Press as well as two anonymous reviewers for taking interest in the story I share herein. Their time, creativity, and faith in this work have enabled it to come to life in this beautiful book.

The People's Spaceship

FIGURE I.1. Space shuttle orbiters *Enterprise* (*left*) and *Discovery* face each other at a ceremony honoring their retirement at the Smithsonian National Air and Space Museum's Steven F. Udvar-Hazy Center in Chantilly, Virginia, on April 19, 2012. Photograph from NASA/Smithsonian National Air and Space Museum.

Introduction

It was a most unusual sight, surreal and sublime all at once. Two space shuttle orbiters faced one another, nose to nose, on a tarmac at the Smithsonian National Air and Space Museum's Steven F. Udvar-Hazy Center, adjacent to Dulles International Airport in Chantilly, Virginia (see fig. I.1). One of the vehicles, *Enterprise*, was bright white against the nearby green foliage and the cloud-dotted, endless blue sky of that warm April day in 2012. After the National Aeronautics and Space Administration (NASA) had used the orbiter for atmospheric drop tests in the late 1970s, *Enterprise* led a sheltered existence as a tourist attraction before retiring into its own wing at the Smithsonian facility. The spacecraft facing it, *Discovery*, was faded and showed signs of wear, bearing the markings it had acquired as NASA's most active orbiter. One year earlier, it had completed the last of its thirty-nine missions to space, just as NASA closed down the shuttle program to free up funds for new human space flight initiatives. For just a few hours, the shuttles stood in this unique configuration. *Enterprise* had been pulled from the Udvar-Hazy Center, and the orbiter would soon journey, first strapped to the topside of a Boeing 747 and then by barge, to a new home in New York City's Intrepid Sea, Air and

Space Museum. The same 747 had carried *Discovery* from Kennedy Space Center in Florida two days earlier, and its cargo would retire by day's end into the hangar *Enterprise* had once occupied.

NASA and Smithsonian personnel, members of Congress, White House officials, astronauts, aerospace industry representatives, and interested individuals from the general public flocked to witness this rare changing of the guard. It was part of NASA's plan to find forever homes for its four decommissioned shuttle orbiters. A military band played patriotic tunes and bystanders waved American flags. Before the ceremony began, renowned opera singer Denyce Graves led the crowd in a moving rendition of "The Star-Spangled Banner." Against this backdrop, dignitaries offered fond words about *Discovery*'s accomplishments. NASA administrator Charles F. Bolden Jr., a former shuttle commander, recalled how "the space shuttle program gave this country many firsts and many proud moments."[1] Looking back over the program's forty-year history, Bolden celebrated the vital role the shuttle had played in deploying and repairing the Hubble Space Telescope and constructing the International Space Station. It allowed people to learn to live and work in space, he said, and motivated future generations of space explorers.

John R. "Jack" Dailey, director of the National Air and Space Museum and once a NASA associate deputy administrator, offered a more curious statement in his brief speech honoring the space shuttle. Like Bolden, he connected the shuttle with the notion of national pride. But Dailey focused momentarily not on the venerable spaceships behind him but on the enthusiastic crowd in his midst. "For every major milestone in space history," he said, "Americans have participated in the excitement, pride, and optimism of the occasion."[2] Indeed, the crowd on hand that day gave validity to Dailey's claim. So too did the multitudes who had set their sights on the skies to catch a glimpse of *Discovery* riding atop the 747 when it approached and circled the Washington, DC, area before landing at Dulles. Since the start of shuttle missions in 1981, Americans had cheered on NASA and the astronaut crews at launches and homecomings of the magnificent flying machine that no other nation in the world could boast. And millions had done the same with human space flight missions before the shuttle debuted, captivated by the landing of three of their

FIGURE I.2. Apollo 11 astronauts ride down the streets of New York City to cheering crowds following their return to Earth. Photograph from NASA.

countrymen on the moon's surface in 1969 (see fig. I.2). NASA is known for its strong commitment to sending people on journeys beyond Earth, and for sharing those profound experiences with ground-bound publics.[3] Because of this, its human space flight efforts have ranked among the most highly visible undertakings of the US government. These sublime ventures into space have astounded people across the United States—and the world over—and have taken their place as a widely recognized American cultural trope.

Dailey's words acknowledged that even those who were not immediately connected with NASA had a place in the storied history of the nation's human space flight program. But their role, according to this characterization, was a passive and reactive one: to observe these spectacles, celebrate them, and feel inspired by the achievements made on their behalf. Indeed, historians and political scientists have typically recognized NASA officials, large aerospace firms, US presidents and other White House officials, and members of Congress as the architects of the American space program. Those outside this sphere typically show up only in accounts explaining that the agency put its feats on display to project a robust US image to people around the globe during the space race with the Soviet Union and to "sell" its Mercury, Gemini, and Apollo human space projects to American citizens who footed the bill.[4]

Yet just ten years after Apollo 11 landed on the moon, an article in *Parade* magazine describing NASA's plans for its new major human flight initiative, the space shuttle, presented a vastly different relationship between NASA and those outside of the government-industry nexus of space program developers. The article noted that the new space vehicle would provide "the first opportunity the public has had to get involved personally in a NASA project."[5] NASA associate deputy administrator Ann Bradley echoed that claim in a 1984 memo. The vehicle's promise of providing routine and reliable access to space to reasonably healthy people with basic training meant that "no development has opened a greater prospect for direct citizen involvement in space flight than the Space Shuttle."[6] According to Hans Mark, NASA deputy administrator when the first shuttle missions began, "the Shuttle opened the door for a vast broadening of the human experience in space."[7] Looking back on the vehicle's legacy, former shuttle manager Wayne Hale elegantly summarized it: "If the intent was to transform space and the opening of the frontier to more people, the shuttle accomplished this. . . . The shuttle truly became *the people's spaceship.*"[8]

What a contrasting perspective these statements offer when compared to characterizations of NASA's public relations activities during the Apollo era! While NASA never abandoned its determination to share the spectacle of human space flight widely, the agency approached public engagement with the shuttle in some new and different ways. Indeed, sustaining the shuttle prompted NASA to rethink how to involve people from across the globe, particularly in an era when other nations were developing capabilities to send humans and cargo to space. But above all, NASA poured tremendous energy into transforming its connections with the American citizenry, whose engagement the agency regarded as paramount to the shuttle's viability.[9] This book tells the story of why and how the agency aimed to involve them as it transitioned from the Apollo period of the 1960s and early 1970s to the space shuttle era that would span the next four decades.[10] It casts a fresh light on the connections between NASA's human space flight initiatives and its public engagement activities, showing how Americans beyond the sphere of government and industry space program players figured in the shuttle program.

Indeed, characterizing the role of American citizens in human space flight solely as celebrants of NASA's achievements does not account fully for their significance in sustaining the endeavor. Doing so leaves us with an impoverished picture of efforts within NASA, even if imperfect, to reconfigure the agency's relationships with its constituents as it sought to move forward with human space activity after Apollo 11 landed astronauts on the moon. NASA saw as crucial to the legitimacy and national support of future human space flight the need to shift from regarding the American citizenry as a single body of unquestioning supporters to one comprising individuals and groups with distinct values, needs, interests, and capabilities and for whom the agency would strive to make the initiative accessible and meaningful. The viability and legacy of the space shuttle in large part depended on NASA's willingness and ability to regard the nation's people not just as potential advocates but as resources essential to the enterprise, even though agency officials constantly struggled with preserving this stance.

Evolution of NASA's Public Engagement Approaches

The commonly perceived connection between NASA and American citizens as witnesses to human space launches is rooted in the agency's mid-twentieth-century origins. After World War II, President Franklin D. Roosevelt's science advisor Vannevar Bush promoted the idea that the results of government-funded scientific research and development activities would ultimately serve the nation's people.[11] In the United States and around the world, government institutions began to consider how their choices to pursue particular science and technology projects could help achieve their visions of desirable futures for their nations, adopting and promoting what Sheila Jasanoff and Sang-Hyun Kim have called *sociotechnical imaginaries* to propagate their ideals.[12] According to political scientist Yaron Ezrahi, conducting technoscientific activities in the open and with claimed commitments to serving the public interest allowed institutions in liberal democracies to gain the seeming approbation of their citizens—whom Ezrahi termed an *attestive public*—as a means to legitimize their actions and authority.[13]

NASA was founded and molded according to these principles. President Dwight Eisenhower responded to the Soviet Union's deployment into Earth orbit of a basketball-sized sphere called *Sputnik* in 1957 by establishing NASA and authorizing it to embark on a program of space research. The White House and congressional drafters of the agency's originating legislation, the National Aeronautics and Space Act of 1958, believed that attaining legitimacy of their vision of America made stronger via a national space program depended on ensuring both global and national public awareness of the agency's activities. Accordingly, the legislation mandated that NASA "provide for the widest practicable and appropriate dissemination of information concerning its activities and the results thereof."[14]

This seminal phrase guided NASA's relationship with those outside the government and the aerospace industry from the start. NASA personnel committed to using all available communications modes to showcase one successful advance in human space flight after another throughout the 1960s in pursuit of an eventual human moonshot—the ultimate display of the nation's prowess. The agency forged strong relations with the news media, welcomed public visitors to NASA facilities and launches, and conducted extensive public outreach and education campaigns. Like the Space Act's authors, NASA's public affairs officials believed that making information about space flight widely available could enhance the United States' image abroad and also garner Americans' appreciation of NASA's activities in support of the vision of national might. NASA officials also recognized that having the backing of the nation's people could serve as an endorsement and reminder of the agency's value to the government officials they elected and who held the treasury's purse strings.

During the 1960s and 1970s, it became clear to government institutions that many Americans would not accept unconditionally the legitimacy of a particular sociotechnical imaginary, policy, or program direction advanced by expert and elite decision-makers. By that time, the use of chemical weapons, nuclear reactors, pesticides, and other contentious creations of research and development activities revealed that science and technology did not de facto benefit all segments of society or the environment. Social activists and scholars began to question the propriety of giving scientific experts and technocrats unchecked authority, with many demanding that all people be given the ability to participate in deciding the rightful place of science and technology within society.[15]

Though having managed to hover above the fray as it fulfilled President John F. Kennedy's national imperative, NASA found itself subjected to societal pressures by the late 1960s. While millions of Americans got caught up in the excitement of the space race, public approval of government expenditures to send people into space was far from unanimous even at the zenith of the Apollo program.[16] Social and economic turmoil at home and a war in Vietnam were taking a toll on American lives and financial resources, leading many citizens to oppose space flight as a quest without clear benefits; meanwhile, the astronauts' homogeneity as white, male, military test pilots seemed out of touch with contemporary public pursuits of civil rights and equal opportunity. Staunch congressional Democrats agreed, and some called for an end to human space flight activities. Put simply, although Apollo 11's lunar landing had created a worldwide sensation, the technological triumph did not ensure widespread acceptance at home. As NASA entered the 1970s, neither the Cold War–era imaginary for human space flight as a pursuit to enhance the nation's global posture nor the idea of the citizenry as an attestive public seemed to hold.

NASA's top leadership was nonetheless enamored of the dream of space travel and saw the human program as essential to the agency's identity and reason for being. Securing political approval for an Apollo follow-on program, however, required a wholesale change in NASA's expectations. While NASA advocated for an ambitious Earth-orbiting space station and crewed trips to Mars, President Richard Nixon proved willing to invest only modestly in human space flight. As this fact became clear, NASA officials pressed for funds to develop just one element of their grand plan: the space shuttle. After originally envisioning the shuttle as a ferry service for astronauts, experiments, and supplies between Earth and the space station, they set to building a case for the shuttle as a standalone program.

NASA's willingness to accommodate military payloads had a significant role in obtaining political support for the new vehicle. But legitimizing the shuttle also entailed finding a completely different sociotechnical imaginary fitting of the new era. While many Americans preferred federal investment in pursuits that would improve citizens' lives, others remained eager to see the nation continue to launch astronauts, and some aspired to be those space travelers. NASA deftly negotiated these varied concerns by constructing the shuttle as a utilitarian, democratic technology.[17] NASA officials conceptualized the reusable new vehicle as a significant departure from the Mercury, Gemini, and Apollo space capsules in both looks and purpose. The shuttle would accommodate seven flyers at a time and operate as a "space truck" that would carry experiments and satellites into space that would benefit people and businesses on Earth. NASA contended that its reusability and projected ability to launch several dozen times annually would save the nation billions of dollars over expendable rockets used only once, while also supporting a multitude of industries and millions of jobs.

Taken together, these arguments allowed NASA to secure approval of the shuttle in 1972 and came to make up NASA's guiding imaginary for the shuttle's role and relationship with the American people.[18] They also signaled that NASA would need to reconsider not only the value of human space flight to citizens but also the instrumentality of those citizens to the human space flight effort and its modes of connecting with them. This book explains how NASA's new vision for human space flight and its changed outlook on public engagement shaped one another throughout the shuttle program.

Chapter 1 opens with an explanation of the origins of NASA's commitment to human space flight and its original quest for an attestive American public. The agency faced substantial challenges in gaining support for a post-Apollo human program, and in chapter 2 I show how officials secured approval for the space shuttle and built a new sociotechnical imaginary to legitimize the project. The next several chapters delineate the myriad ways in which NASA sought to engage various publics with the shuttle program. In chapter 3 I explain how NASA officials aimed to enlist specific groups as supporters—from community business leaders to *Star Trek* fans—by interacting directly with them and tailoring messages and opportunities to satisfy their interests. NASA also continued to engage different publics through display as it did with its human missions up through the early 1970s. In chapter 4 I reveal that the agency advanced its imaginary for the shuttle by leveraging the shuttle's unique physical attributes and available communications technologies to aid all Americans in feeling connected to the vehicle.

The shuttle program's success, however, depended on more than persuasive arguments and enhanced communications approaches. Rather, it would be crucial to *realize* the vision of a democratic human space flight program. In chapter 5 I show how NASA sought to prove the shuttle's utility by inviting involvement by an eclectic variety of people. Human space flight officials suddenly found themselves soliciting satellite and experiment customers, from major companies to individual citizens. In chapter 6 I explore how the agency expanded the human space flight ranks to include new flyers aboard the new vehicle. From broadening the demographic and professional diversity of its astronaut corps to initiating a program to welcome flights by private citizens, NASA aimed to ensure that those who rode on the shuttle resembled America.

Just a few years after shuttle operations began, it was evident that NASA's relations with American citizens were markedly different than they were two decades prior. According to many indicators, the agency's approach bore fruit. Across the nation, the shuttle instilled a sense of national pride and common ownership. Even so, democratizing the shuttle was not easy for NASA: fostering the vehicle's accessibility required the agency at times to navigate criticisms from the media, the commercial launch industry, and even some of NASA's own scientists and astronauts. Moreover, the shuttle's complexity kept flight rates to just a fraction of what the vehicle's proponents had anticipated.

As I reveal in chapter 7, NASA's commitment to opening the shuttle to more public involvement encountered even greater difficulties beginning in 1986. That year, the space shuttle *Challenger* launch disaster killed seven astronauts, including NASA's first "citizen in space," teacher Christa McAuliffe. The accident also ripped asunder NASA's imaginary of a vehicle accessible to and capable of serving the nation's citizens as it revealed a technology fraught with risks and tarnished NASA's credibility and image of competency in the eyes of the media and many elected officials. Although NASA officials had for years regarded direct involvement of broad segments of the public as crucial to the shuttle's viability and legitimacy, they began to temper their enthusiasm for inviting new shuttle users and flyers as they revisited future direction in human space flight. I show in chapter 8 that by the time of the *Columbia* accident in 2003, NASA had moved away from adhering to the sociotechnical imaginary of a democratic shuttle yet celebrated the vehicle's legacy as the people's spaceship as the orbiters completed service and were sent to new homes in museums around the country.

New Perspectives on Public Engagement and Space History

NASA never described its interactions with its various publics during the early shuttle era as "public engagement," as this term entered the popular and scholarly lexicon later. Even so, as chronicled in this book, NASA's experience with the space shuttle reveals that many factors drive and constrain the visions, abilities, and approaches of a technoscientific agency within a democratic government to engage with its constituents. Democratic governments establish such agencies to contribute to a nation's welfare, and agency leaders must make decisions on an ongoing basis about the development, use, control, or stewardship of technologies. In doing so, they recognize the need to demonstrate value and accountability to the publics that, at least indirectly, sustain them.

NASA's efforts to involve disparate publics with the shuttle shows that it is incredibly challenging for a technoscientific agency to achieve these aims. Many NASA officials tried to make the shuttle as open and inclusive as they could but encountered resistance from some inside the agency or with longstanding relationships to the space program when they sought to invite others to participate, given that the shuttle was a limited resource. The agency thus had to prioritize how it served various publics through the shuttle while remaining cognizant of how its choices would be received by congressional stakeholders who directly determined NASA funding levels. Indeed, when democratic government institutions are in principle obligated to "think" in the aggregate and serve millions of people, whose opinions are far from harmonious or evenly valued, it is virtually impossible for them to satisfy everyone or pursue a consensus direction. In some senses, it is rather remarkable that NASA was at all able to expand and maintain opportunities for more citizens to participate in the shuttle program.

This book also rebalances historical understanding of NASA's public relations efforts with human space flight. Mapping the evolution of public relations during the shuttle era reveals the internal and external considerations that influenced NASA's complex relationships with various publics and its public engagement strategy for four decades. It shows who mattered throughout the program and why. NASA operated for its first decade as if it had one big, attestive public—an undifferentiated citizenry who, dazzled by the spectacle and patriotism of space flight, would support it. But when overwhelming support did not materialize even after Apollo 11's landing on the moon in 1969, the agency sought to involve segments of society that could help to make the shuttle a success substantively, culturally, and politically. To that end, this work is a departure from space histories that cast the American public in a passive and consumptive role.

This examination of NASA's public engagement approaches during the shuttle era expands the historical narrative of the American space program in still other ways. While the agency's outreach efforts during the shuttle era undoubtedly incorporated the support-seeking and image-building pursuits that were rife during the Apollo days, public engagement strategies around the shuttle can be seen more charitably as attempts to remain accountable to the American people in ways meaningful and suited to the times. Beginning in the early 1970s, NASA's leadership believed that the agency could deliver on its sociotechnical vision and attain legitimacy for the shuttle only by enrolling Americans in the shuttle program, and this required engaging more people according to their varied interests. NASA's external engagement approaches served many segments of society while they were supporting the agency's continued quest for public and political approval. This symbiotic relationship meant that the shuttle was, in at least one respect, truly shared.

Some pundits have considered the shuttle to be a mistake for the American space program. Political scientist John Logsdon has called the shuttle a policy failure, stating that budget constraints imposed by the Nixon administration pushed NASA down a path of making overzealous promises that the shuttle would reduce the cost of space flight and become self-sustaining financially, and that the White House's choice of NASA's shuttle design precluded other space program directions.[19] Former NASA administrator Michael Griffin lamented shortly after taking office in 2005 that the shuttle's design was "extremely aggressive and just barely possible" and had left the nation with a flawed human space flight system.[20] While the shuttle had its share of imperfections, its inclusion of and service to diverse publics in ways that earlier human space flight initiatives had not provides yet another figure of merit by which the program's evolution ought to be understood and its success judged.

Those within NASA who saw the value of a democratic space shuttle program overcame hurdles as best they could to create opportunities for more people to participate in space flight activities. Looking at NASA's aims and approaches to engage disparate publics with the shuttle in ways meaningful to them over the vehicle's four decades can help the agency determine how it can best evolve its program plans, public engagement mechanisms, and performance measures to connect with various publics as it pursues human exploration of the moon and beyond. The public relations issues NASA encountered during the long shuttle era are indicative of at least some of the challenges the agency will need to address.

1

Building a National Vision and Public for Space Flight

The United States has launched more people into space than any other country, and Americans have made up more than half of all human space travelers throughout history. Some 45 percent of NASA's current budget is dedicated to vehicles, systems, and research to support human space flight. However, an American space program focused heavily on human explorers was not inevitable. Nor was it guaranteed that astronauts would work as civilians, conducting their activities in an open fashion. The nation's commitment to human space exploration was driven by technological, political, and cultural developments that shaped one another over time.[1] US government officials decided to put space initiatives on display with the expectation that doing so would allow them to be better understood and appreciated by American citizens and revered by the global public. In this chapter I outline the origins of the commitment to a US civilian human space flight program, highlighting how national and NASA leaders perceived an attestive American public as important to their success and strived to obtain widespread support by conveying an imaginary linking space activity with national prestige.

The dream of soaring into space had captured the imagination of people throughout the world for millennia. As early as the fourth century BC, the Indian epic poem *Mahābhārata* described flights of vehicles across the cosmos. Literature about travel to the moon and planets began appearing with increasing frequency as the discoveries of Galileo Galilei, Johannes Kepler, and other Europeans emerged in the seventeenth century, including books such as Francis Godwin's *Man in the Moone*. Jules Verne's works on space travel influenced Robert Goddard, an American; Hermann Oberth, a German; and Konstantin Tsiolkovsky, a Russian. Working separately, these three established and proved the principles of modern rocketry in the first part of the twentieth century. Their research in turn laid the groundwork for Germany to develop the world's first long-range missile, the V-2, which the Nazi regime used to launch offensive attacks on Allied European nations during World War II.

Wartime developments led the US government to take an interest in rocketry and, ultimately, human space travel. During the war, the National Advisory Committee for Aeronautics (NACA), which would become the organizational nucleus of NASA, began to investigate the aerodynamic properties of small rockets that could serve as high-speed guided missiles under the leadership of Robert Gilruth, an engineer at the Langley Research Center in Hampton, Virginia. Further south, at Fort Bliss, Texas, and eventually at Redstone Arsenal in Huntsville, Alabama, the V-2's mastermind, German rocket scientist Wernher von Braun, started working with the US Army to develop ballistic missiles. Von Braun had surrendered to the United States following the war, eager to continue his rocketry work and apply it to fulfill his personal dream of creating more perfect societies in colonies across the solar system. He sought political and public support for a space exploration program that achieved this aim beginning with artificial satellite launches and followed by orbital human flights, a human-tended space station, and human visits to the moon and planets. Von Braun collaborated with *Collier's* on a series of magazine articles and with Walt Disney to make television appearances to promote his vision, conveying through these media that the technology for space travel was within reach. He also preyed upon growing American fears of the military capabilities of communist countries while highlighting an opportunity for global leadership, noting that a spacecraft "could be either the greatest force for peace ever devised, or one of the most terrible weapons of war—depending on who makes and controls it."[2]

Von Braun's tactics were effective in government circles: NACA leadership recognized the merits of von Braun's vision, and Gilruth helped to create an interagency board aimed at attaining human space flight capabilities in 1952.[3] The Soviet Union's launch of *Sputnik* on October 4, 1957, and its launch of *Sputnik II* carrying a dog named Laika one month later moved the United States ever closer to a commitment to develop a human space flight program. NACA began formulating plans for human activity in space in early 1958. The US Air Force had also been contemplating the merits of a human space flight program and intensified its own studies.

Even despite the Soviet success, President Dwight Eisenhower wanted the United States to take a measured approach to space activity, preferring to concentrate on satellite applications and international scientific cooperation above human space flight. He wanted the United States to carry out its space projects in an open, peaceful manner, in contrast with the closed, secretive Soviet program. Eisenhower also had misgivings about giving the military control of space activities, given the rivalries that it could engender among the services as well as the negative impact it could have on worldwide opinion of the United States. Consequently, he established a civilian space agency, the National Aeronautics and Space Administration (NASA), which subsumed and expanded on the activities of NACA. The new space agency began operations on October 1, 1958. Eisenhower urged NASA to pursue a portfolio of scientific and applications research to yield long-term value to the nation's citizens and "for the benefit of all mankind"—direction that was incorporated into the agency's originating legislation, the National Aeronautics and Space Act.[4]

A blue-ribbon panel of academic social scientists assembled by NASA's first administrator, T. Keith Glennan, agreed that emphasis on projects with practical applications would give the agency the greatest public appeal. The new agency thus built a long-range plan in 1959 that included a host of meteorological, communications, and space science satellites and interplanetary probes that would advance scientific aims, contribute to numerous civilian and military applications, and support the national economy.[5] Nevertheless, Glennan, along with leaders from the United States Information Agency, the State Department, and the National Security Council, subscribed to the views of von Braun and Gilruth that a human space flight program would create public excitement and build national prestige.[6] Consequently, the agency included in its long-range plan a human circumnavigation of the moon and construction of a space station targeted for after 1965 and human flights to the lunar surface beyond 1970.[7]

While Eisenhower hesitated to prioritize space stations and trips to the moon, the president did see value in demonstrating the feasibility of human space flight. Intelligence reports indicated that the Soviets were working toward launching a man into space. Eisenhower assigned NASA responsibility for developing and executing a human space flight program and directed the air force to cease its own efforts in this arena and instead collaborate with the new agency. The objectives of the NASA-led initiative, dubbed Project Mercury, were to put a human in orbit, understand the associated technical requirements and medical effects, and bring the space traveler back safely. But as astronaut Alan Shepard was preparing to launch aboard the first suborbital flight toward meeting Project Mercury's goals, geopolitical events would lead the United States to amplify its commitment to human space activity and adopt a sociotechnical imaginary that NASA would use to justify its direction to the American public for the next several years.

Forging a National Imperative

Like his predecessor in the Oval Office, President John F. Kennedy was initially unconvinced that human space flight ought to be the focus of the United States' civilian space effort. At the start of his administration, he, too, preferred a methodical and balanced approach to NASA's projects. Kennedy also desired to cooperate with the world's nations, including the Soviet Union, on an array of space activities. But in April 1961 two developments prompted the new president to pursue a more aggressive, competitive human flight program. On April 12, the Soviet Union's launch of Yuri Gagarin, before the first Mercury flight, propelled the communist nation ahead of the United States in space feats. Five days later, the failed Bay of Pigs invasion dealt a deep blow to the Kennedy administration and to the

United States' image abroad. In reaction to these events, Kennedy asked Vice President Lyndon Johnson for options to outdo the Soviets in space.

Johnson worked with members of the National Aeronautics and Space Council to consider alternatives. The group had been established in the National Aeronautics and Space Act to advise the president on space policy matters; at the time, it was chaired by the vice president and included the secretary of state, the secretary of defense, the NASA administrator, and the chairman of the Atomic Energy Commission. On April 28, 1961, Johnson delivered a report to the president. Asserting that countries around the globe increasingly identified "dramatic accomplishments in space" as "a major indicator of world leadership," the report claimed that with a strong commitment, the United States "could conceivably be first" in a human circumnavigation of the moon or a human landing on the lunar surface by 1966 or 1967.[8] Kennedy's new NASA administrator, James Webb, and Defense Secretary Robert McNamara recommended an across-the-board acceleration of the US civil space program while highlighting the importance of sending humans to the Moon, stating that "it is man, not merely machines, in space that captures the imagination of the world."[9] The national prestige accrued from such a success, they argued, would be substantial, even if the endeavor had limited scientific, commercial, or military value.[10]

All the while, the Kennedy administration strived to engage the Soviet government in cooperative space projects, including a joint lunar mission. The Soviets, however, rebuffed the suggestion without a US commitment toward disarmament. Kennedy pressed forward by accepting the recommendation and announced his intent before a joint session of Congress on May 25, 1961, for NASA to land humans on the moon "before this decade is out" (see fig. 1.1).[11] Framing human space flight as a great adventure critical to winning "the battle that is now going on around the world between freedom and tyranny," Kennedy's appeal married a far-reaching technological goal with a vision of American leadership and exceptionalism among the world's nations. In doing so, the president solidified NASA's guiding imaginary for human space flight. Kennedy's decision, which Congress supported initially with more than half a billion dollars as a supplement to NASA's 1961 budget, reversed Eisenhower's strategy of

FIGURE 1.1. President John F. Kennedy's May 25, 1961, speech to Congress established the sociotechnical imaginary of an America made stronger on the global stage by human space activity. Photograph from NASA.

avoiding a space program focused on technological showmanship at the expense of other elements. The enormous influx of resources for human space flight helped to embed a dominant culture within NASA marked by zeal for ambitious engineering projects centered on the human program.

Not all who were close to NASA favored the agency's adoption of this vision for space flight. Funding continued throughout the 1960s to support NASA's nascent initiatives to explore the planets with robotic sentinels and to observe the sun and universe with telescopes placed in orbit, but these projects received far fewer resources than the Apollo flight systems projects. Many scientists protested NASA's priorities, expressing worry that Apollo project managers would neglect scientific opportunities for the lunar missions even when the Space Act called for NASA to enable the science community to make observations and measurements from space. The National Research Council's Space Science Board raised these concerns during a 1962 review of NASA's science programs.[12] The following year, Philip Abelson, editor of the journal *Science*, argued in an editorial and later in testimony before the Senate that robotic probes would prove a far better way to conduct scientific research than through human missions.[13]

Nevertheless, Kennedy's presidency was extinguished prematurely by an assassin's bullet, and the Johnson administration continued the nation's commitment to supporting Apollo. At its funding peak in 1966, the program consumed almost $3 billion, or nearly two-thirds of NASA's total budget and some 2.5 percent of all US federal spending for that year.[14] While academic space scientists expressed displeasure with the funding flow, others benefited handsomely. Apollo facilitated a massive mobilization of aerospace industry contractors dedicated to the lunar landing program as NASA sought to extend its capabilities and workforce through strategic partnerships with those who had expertise and experience developing large defense systems. Contracting with industry also spread funds around the country. Sure enough, members of Congress from states such as Texas and Alabama, where a sizable amount of space program development work took place, became staunch supporters of NASA and Apollo.

Eisenhower had fretted about the implications of technocracy and a powerful military-government-industrial complex, cautioning in his 1961 farewell address that "public policy could itself become the captive of a scientific-technological elite."[15] Bearing this out, it was the exclusive nexus of NASA, the aerospace industry, and Apollo's supporters in the Kennedy and Johnson administrations and Congress that championed the pursuit of human space flight as a national imperative throughout the 1960s. But they recognized that their success depended not just on overcoming technological hurdles but also on another critical task: getting the American citizenry to buy into their sociotechnical imaginary for space flight.

Seeking an Attestive American Public

Almost all US federal government agencies have public information or consumer affairs offices that keep Americans and global audiences apprised of their activities and services. Language about communicating with the public was incorporated into a few agencies' originating legislation, NASA's included, and an examination of the laws and their histories provides insights into the motivations for encouraging sharing of information. In the post–World War II era, US political leaders recognized, for one, that knowledge exchange could help to advance science. The Atomic Energy Commission was "permitted and encouraged" by the Atomic Energy Act of 1946 to disseminate scientific and technical information relating to atomic energy "so as to provide that free interchange of ideas and criticisms which is essential to scientific progress."[16] Likewise, in envisioning a civilian space agency, Eisenhower wanted to ensure open and free exchange of data to facilitate cooperation among domestic and international scientists to use space peacefully and to produce societal benefits. In a message to Congress, the president stated that the new agency's originating legislation should consider "matters related to dissemination of the data collected" through space activities.[17]

Members of Congress who penned the Space Act agreed on the importance of an open space program for scientific cooperation. The final version of the Space Act thus directed NASA "to provide for the widest practicable and appropriate dissemination of information concerning its activities and the results thereof."[18] But the Space Act's authors saw merit in including this directive for still other reasons. As the House and Senate heard the views of dozens of witnesses in order to prepare their respective versions of the legislation, members on both sides came to recognize that conducting space projects in the open would help the United States demonstrate global leadership, promote democratic values abroad, and create a national image distinct from, and preferable to, the Soviet Union's communist political regime and socioeconomic system. If people around the world could watch, learn about, and revere American space successes, then the aim of affirming the United States as a technoscientific authority and the premier spacefaring nation, they believed, would be fulfilled.

The space program's progenitors had global audiences in mind when issuing the directive to NASA to share information, but they also wanted to be sure that the American public learned of the space program's accomplishments and value. The Space Act directed that, classified and proprietary data notwithstanding, "information obtained or developed by the Administrator in the performance of his functions under this Act shall be made available for public inspection." The report accompanying the House version of the bill indicated that the statement's intent was "to let the people know all the facts."[19] This language implied a sense of obligation to be transparent with Americans as well as an acknowledgment that the public had a right to know what the new space agency was achieving with their tax dollars.

There was, however, a persuasive aspect to the Space Act's public communication clauses as well. The idea of space flight was not new to Americans when Eisenhower created NASA. It had been popularized by science fiction writers such as Arthur C. Clarke, Isaac Asimov, and Robert A. Heinlein, and during the 1950s von Braun had promoted it as a distinct possibility in his *Collier's* articles and television programs with Disney. By the time of *Sputnik*'s 1957 launch, a majority of people believed that humans would set foot on the moon within a quarter-century.[20] Although many people were willing to part with some pocket change to enjoy a space-related film or book, opinion research showed that no identifiable public constituency strongly supported a government-funded humans-in-space program. Analysis completed by Donald Michael, a social psychologist with whom NASA contracted to study space flight's social implications, indicated that, rather than worrying about *Sputnik*, "for many people everywhere, their own affairs, Little Rock,"—an allusion to desegregation—"and the World Series took precedence over the Soviet leap into space." Michael concluded that "there is no good reason to believe that there will be strong pressure from the public for effort and expenditures in this area, *unless very special efforts are made to elicit it.*"[21]

Members of the House of Representatives who drafted the Space Act shared this concern. The House report accompanying the Space Act noted that "only with thoroughgoing public understanding can the necessary support be marshaled to make it a success."[22] It was not the first time legislators had sought Americans' support for an initiative they believed was in the national interest to pursue. A year after the Atomic Energy Act was signed into law, Congress amended it to include "public understanding" as a reason for disseminating information after an advisory committee questioned the nation's willingness to accept the use of atomic energy for peaceful purposes. In the case of the Space Act, the authors assumed that making information available could alleviate any doubts Americans had about the space program and cultivate an attestive public that stood behind, and justified, major national investments in space activities.

The notion that a better-informed population would lend its support to research and development projects—today dubbed the "deficit model" of science communication—resonated with NASA officials as well. Consequently, the agency embraced with zeal the mandate to disseminate information and used it to define its interactions with the American people outside of policy makers, academic scientists, and the aerospace industry. Glennan established the Office of Public Information, an offshoot of the modest public relations office of the agency's predecessor, NACA. While the 1919 Anti-lobbying Act bans agencies from using their budgets, furnished by citizens' tax dollars, to persuade members of the public and Congress to support them and the programs they conduct, NASA administrators and public affairs officials set out to garner public understanding and appreciation, especially of human space flight.[23]

Indeed, NASA officials recognized a deep need to create constituencies for the human space flight program. One month following Kennedy's appeal to Congress to send humans to the moon, only 42 percent of Americans favored the nation's commitment of tens of billions of dollars for Apollo.[24] A few years later, sociologist Amitai Etzioni would write *The Moon-doggle: Domestic and International Implications of the Space Race*, a book blasting what Etzioni believed to be misplaced US national priorities.[25] In May 1961 deputy administrator Hugh Dryden testified before the Senate Appropriations Committee that NASA needed to help the public understand the Apollo project for them to accept it. Dryden stated, "Now whether Americans want to accept that for scientific reasons, prestige reasons, propaganda reasons, or all three, is something we up here must decide."[26] Brian Duff, director of NASA's public affairs division in the 1960s, explained that NASA made a point of sharing information because its activities "had to have public understanding and public support."[27]

While NASA sought to sell the American and global publics on Kennedy's imaginary of a nation made stronger by human space flight, the agency mixed its championing of the national imperative with its own preservation. As Charles Biggs, a public affairs officer at the Manned Spacecraft Center during the 1960s, intimated, NASA walked a fine line between serving the public and promoting itself in conducting the open program: "We never did anything improper or inappropriate or illegal, but we would bend every rule that we could to do something if we thought it was good for the agency, we thought it was good for the image, we thought it informed the public. That's what the Space Act is."[28]

Webb, NASA's second administrator, felt the same way. While being understood by the American public could promote accountability, Webb made clear that this was not the reason he valued informing citizens about NASA's work. Rather, he considered their understanding to be critical to legitimizing NASA's activities as planned. Duff recounted Webb attending a NASA program review in which a presenter explained that NASA had an open program out of respect for the public's right to know. According to Duff, Webb refuted the comment, stating, "We have an open program because it is good public policy to have an open program. It's good for NASA and it's an effective NASA policy to have an open program. I'm not in the business of the public's right to know. There are others like the attorney general who will take care of that."[29]

Webb believed that NASA's "shareholders" were the American taxpayers and that they were ultimately the people to whom the agency had to answer. But he regarded Congress as the agency's "board of directors" and associated getting Americans to understand and take interest in the space program with garnering political support to fund NASA's programs.[30] In the mid-1960s, the agency had strong backing from the White House and a few key members of Congress, including George Mahon of Texas, who became chair of the House Appropriations Committee in 1964. According to Duff, Webb saw public support as essential nonetheless because "it's the public support that lets these other people give us what they want to give us. It's George Mahon knowing that back in Lubbock, his constituents are quasi-convinced that the space program is a good thing."[31] And, by this logic, marketing the space program could also help win over legislators who opposed the costs of space flight but who might relent if they believed their constituents felt that NASA was serving their interests.

For the most part, however, agency officials and documents denied any semblance of attempts at self-preservation, emphasizing instead a commitment to serving American citizens through the human space flight program and its mandate to disseminate information widely. A 1966 document assembled by NASA's public affairs program stated straightforwardly: "We are not in an image-building business; we are trying to create a program that reflects what NASA is and what it is accomplishing. This is a distinction and a very important one. . . . We feel that we have a service to offer."[32] Administrator Thomas O. Paine, who succeeded Webb, also argued for the neutrality of NASA's information sharing efforts, apprising Congress that: "In NASA we do not translate *interest in* as *support for*. We only acknowledge the interest and try to supply goods and services on a reactive basis. And it is extremely important to appreciate the fact that these goods and services are factual, not self-serving and not message-bearing."[33]

Regardless of stated motivation, NASA public affairs officials aimed to provide American citizens access to information, people, and facilities associated with human space flight in a myriad of ways. Much of their focus was on building connections with members of the news media who, in that era, served as the US government's gateway to communicating with the American public. NASA charted new ground, forming what former Kennedy Space Center public information chief John W. "Jack" King called "adversarial" but "friendly" relationships with journalists.[34] At no time was this more salient than when in 1959 the agency signed a deal with *Life* granting the magazine exclusive access to the Mercury astronauts' personal stories. In 1963 NASA offered a grant to Columbia University's journalism school to study and improve how the agency could communicate its accomplishments to the public—although Columbia rejected the funding when media critics charged that the project focused on improving NASA's image.[35]

NASA made itself broadly available to media outlets as its human space flight efforts progressed. The agency initially released information about launch plans only a short time prior to a launch and provided little status on vehicles in flight, embracing a policy to "do first and talk later" and averring that "our record should be built on the solid foundation of

FIGURE 1.2. NASA holds a press conference on March 24, 1965, the day following Gemini 3's three-orbit mission. Seated at the table from left to right: Kurt H. Debus, Kennedy Space Center director; Christopher C. Kraft Jr., Manned Spacecraft Center assistant director for flight operations; John Young, Gemini 3 pilot; Virgil I. "Gus" Grissom, Gemini 3 command pilot; Robert R. Gilruth, Manned Spacecraft Center director; Robert Seamans, NASA associate administrator; and Julian Scheer, NASA assistant administrator for public affairs. Photograph from NASA.

achievement."[36] Julian Scheer, however, expanded NASA's accommodations for journalists, arranging news centers and press sites at launches and granting them access on a pool basis to mission control rooms and astronaut recovery ships (see fig. 1.2). Scheer also pressed Apollo's technical teams to include television cameras on the lunar landing missions—items not on the original mission requirements list—and to allow live media coverage of transmissions between mission control personnel and astronauts aboard spacecraft. Many astronauts and mission officials resisted on grounds that such openness might reflect poorly on NASA and the nation if, say, a problem arose during flight or an astronaut let foul language slip on live air. However, top NASA officials, including Paine and Gilruth—who had become director of the Manned Spacecraft Center in Houston—almost always overruled the objections in favor of preserving the open program.

NASA also reached out directly to a wide variety of publics. The administrator, other officials, and astronauts gave speeches to hundreds of professional and civic societies each year. The agency also received thousands of public inquiries, some praising NASA but others critical in nature. In engaging with these audiences, NASA officials strived to articulate the value of human space flight, framing it as a service in the nation's interest. Early arguments swirled around the Cold War imaginary of bolstering the nation's security, maintaining influence in world affairs, enhancing US prestige abroad, and advancing the cause of freedom over the perceived communist threat. As NASA deputy administrator Dryden conveyed in a 1962 speech to the United States Savings and Loan League titled "The Citizen's Stake in Space Exploration," a human lunar landing would not just serve as "a spectacular achievement" but "create for us a space capability second to none—the all-around power to exploit space fully in the national interest whatever that interest may require."[37] NASA officials also connected the initiative with national leadership by invoking traditional American values. Scheer noted in his letter to Mrs. Edward Levine of California that the endeavor would "carry on the best pioneering traditions of the American people."[38] In a response to Mrs. E. G. Hunter of Massachusetts, Scheer stated: "This move out into space is a tremendous human venture and adventure. It is a challenge to the spirit of man that our Nation, with its tradition of leadership and forward moving, cannot fail to accept."[39]

NASA's rhetoric sometimes implied a level of responsibility, participation, choice, and sharing of national interest in human space flight by American citizens when most were not included in the decision-making process. Scheer responded to a letter from one man, Terrence O'Neill, asking about the rationale for a national space program by stating, "The American people have made clear their desire to play a leading role in this exciting new field of human endeavor."[40] A master of invoking an imagined American public committed to space flight to garner support, von

Braun told a public audience in Lexington, Massachusetts, in 1959 that he was informing them about the benefits of space exploration so that each citizen could make "reasoned judgment" and "exercise his privileges as a citizen to achieve objectives which he considers essential to the welfare of his country."[41] Dryden's "Citizen's Stake" speech noted that the masses would have a virtual role in the adventure of space flight: "The success of the National Space Program hinges on the ability of the American people, through their government, their industry, and their privately endowed institutions, to implement many difficult tasks. Not one or two men will make the landing on the moon, but, figuratively, the entire Nation."[42]

The public affairs offices at NASA Headquarters in Washington, DC, and the NASA installations around the country also made provisions to showcase the human space flight program directly to American citizens. Through pamphlets, films, traveling exhibits, a speakers bureau, and other outreach mechanisms, they aimed to educate and inform people about NASA's accomplishments and plans and to enable them to "share freely in the excitement and adventure of this peaceful research and exploration into space."[43] NASA partnered with the Department of Defense to sponsor a large "Space Park" exhibit at the world's fair in New York City in 1964, which Webb considered to be "one of our best opportunities to show manned space flight in its true relation to the other segments of our program and to the development of space pre-eminence for the United States."[44] In 1962 the agency took the *Friendship 7* Mercury capsule piloted by John Glenn to major cities as well as remote locales around the world. During 1970 and 1971 the Apollo 11 capsule and moon rock samples visited all fifty US state capitals. NASA noted that the nationwide tour would "enable the public to share the Apollo 11 adventure."[45]

In addition, NASA's field centers, including the Launch Operations Center on Cape Canaveral in Florida (renamed Kennedy Space Center in 1963) and the Manned Spacecraft Center (renamed Johnson Space Center in 1973), began tours so that the visiting public could see the hardware and facilities that supported the nation's space program (see fig. 1.3). These tours were based on a tradition that began at certain NACA installations to hold open houses. Scheer encouraged NASA officials to invite people from outside of the sphere of traditional space program

FIGURE 1.3. A Kennedy Space Center tour group views a Saturn V rocket on the launchpad.
Photograph from NASA.

developers to the Apollo launches to "educate" and make them "aware" of NASA's activities, stopping short of claiming to seek appreciation and support.[46]

NASA further aimed to educate Americans to "achieve for our country a science-literate citizenry able to understand and act intelligently in the face of many problems emerging from an age of science and technology."[47] Among the agency's efforts was a pilot project with Howard University to provide "space-age understandings" for adults with limited reading ability. But NASA's most notable formal and informal initiatives to create public understanding of space activities reached beyond voters and taxpayers. The launch of *Sputnik* had galvanized science education within government and academic institutions throughout the country, and the agency recognized the importance of appealing to students to become the nation's future scientists and engineers. As Webb stated, "Among the major motivations of the space program is . . . the stimulating effects of this challenging national enterprise on all segments of American society, particularly the young."[48] NASA established an education office to develop materials for teachers to explain space concepts to their classes and sent NASA staff to visit schools and communities with "Spacemobile" vans filled with models and science demonstrations. The agency set up a university program to support higher education as well as efforts to attract early interest in space-related fields and careers among elementary and secondary students.

NASA officials felt encouraged to continue their campaigns to inform and educate the public because of strong positive feedback they received in the media and from space flight enthusiasts. Reporters and members of the public reacted to NASA's presentation of the Mercury astronauts in 1959 with what historian Roger Launius calls a "wave of excitement."[49] Requests flooded in for astronaut appearances at a variety of events. Various media outlets heaped praise on the agency's approach to public engagement. A 1965 article in the *Washington Post* noted that NASA's open information policy "has made every American a participant in its exciting conquests of space." Quoting a West German publication, the story also noted: "The Americans make it possible for the public around the world to participate without reservation in all phases of their exploration of space. . . . [T]he advantage is that the observer feels caught up in the adventure to a greater degree than if he were served up a success with victory fanfares in an all-inclusive communiqué."[50]

The heady volume of requests NASA received from members of the public for information further validated the agency's communications approach. When the Apollo 11 astronauts reached the moon in 1969, more than 60,000 letters flowed into the agency's mail rooms each month, with huge numbers of them addressed to the astronauts. Every day, in NASA's estimation, 50,000 people viewed NASA exhibits, 9,500 requested NASA publications, 10,000 schoolchildren enjoyed Spacemobile demonstrations, and 16,000 watched a NASA-produced film. Demand for Spacemobile visits far exceeded NASA's ability to fulfill all requests. In 1969,

2.6 million people toured NASA facilities. Comment cards collected from Kennedy Space Center tour participants indicated that less than 1 percent of respondents were critical of the space program, with many visitors expressing that the tour made them feel "grateful" or "proud" or showed them the "necessity" of the space program.[51]

Taking stock of this positive response at the height of the Apollo program, Scheer's successor as public affairs director, John Donnelly, reflected in 1971: "Thanks to the candid, responsive public affairs posture it has maintained, particularly with respect to manned space flight, NASA has earned an excellent credibility rating with the press and the public."[52] But looking deeper, opinion polls and other developments belied that claim and cast doubt on NASA's imaginary for space flight. Despite the enthusiasm NASA witnessed, more than half of Americans believed that the government should stop funding Apollo and focus instead on addressing domestic needs and civil unrest. Some questioned the value of human space flight following the tragic deaths of the three Apollo 1 astronauts who perished in a fire on the launchpad during a training mishap in January 1967.[53] Representative William Fitts Ryan of New York took particular umbrage to NASA's handling of the Apollo fire and expressed his conviction that "the public which, at a tremendous cost, is financing NASA's efforts in space, has a right to know how its money is being spent . . . the public has not been allowed to review [NASA's future] plans and to approve of future expenditures."[54]

NASA remained committed to diffusing information about human space flight, though not for the reasons the congressman advocated. Beyond grooming future generations of its workforce, NASA largely sought to create understanding and fulfill and fuel Americans' interest with the aim of gaining an attestive public for human space flight. The agency did all it could to make the nation's voters and taxpayers spectators and fans of its activities. But not everyone was enamored of NASA's cause, and the concerns of human space flight skeptics would become too pronounced for the agency to brush aside. They would factor into choices about human space initiatives—and how to engage the public with them—in the post-Apollo era.

2

Bringing Human Space Flight Closer to Earth

If one thing was clear after the Apollo 11 *Eagle*'s landing on the moon on July 20, 1969, and the astronauts' safe return four days later, it was that the space agency had created an impressive legacy on multiple fronts. Visiting another world had been the stuff of dreams for centuries, and now humans had done it for the first time. It showed the capability of the human mind to think up the seemingly unattainable and craft a technological system to attain it—and for a nation's leaders to see its value and allocate the necessary resources. The achievement proved a major display of American capability and showmanship, power, and prowess in the highly visible "space race" theater of the Cold War with the Soviet Union. As a result, NASA basked in perhaps the biggest public spotlight in human history. Millions had lined the beaches around Cape Kennedy for the moonshot. An estimated six hundred million people around the globe, some 20 percent of the planet's population of the time, tuned in via television to witness Neil Armstrong and Edwin "Buzz" Aldrin take humanity's first steps on an alien world.

NASA's administrator, Thomas O. Paine, was determined to gain President Richard Nixon's approval for the space agency to continue to press the boundaries of human space flight. Two weeks after the Apollo 11 crew's return to Earth, Paine proposed a major post-Apollo human space flight program. The plan called for extending human lunar activity with orbiting and surface bases, a permanently occupied Earth-orbiting space station to be serviced by a reusable "space shuttle," and a twelve-person expedition to the surface of Mars launching in 1981, for a cost of $4–8 billion annually. Paine might have advocated that NASA consider its human space flight work complete with the Apollo lunar landing and instead shift its emphasis. After all, the National Aeronautics and Space Act of 1958 that had given birth to NASA called only generally for NASA's "preservation of the role of the United States as a leader in aeronautical and space science and technology and in the application thereof."[1] The agency had over the previous decade developed capabilities and made advancements in interplanetary exploration missions, weather probes, and satellite communications. Putting more resources into these areas would have sat well with the space science community and liberal members of Congress, who felt that these initiatives had greater scientific and societal value.

But Paine was a human space flight enthusiast—a "true believer" among the agency's legions of human space flight proponents who, as characterized by one former NASA manager, maintained with "almost a religious fervor" that the expansion of human presence into space was critical for the nation and humanity's destiny.[2] The human program formed the heart of the agency's internal culture and had earned it world renown, and Paine wanted to take it further. He also recognized that many of the engineers and program managers in NASA and the aerospace industry who had enabled the Apollo effort would share his desire to enable humanity to reach deeper into space.

While ego and self-interest were clearly at stake in Paine's bid for Mars, he and other NASA officials believed that the agency had served the interests of the American people through the Apollo program and its human flight program predecessors, Mercury and Gemini. Indeed, the national commitment had built up expertise and technical capabilities at NASA facilities and within the aerospace industry. It also allowed the nation to enjoy newfound worldwide prestige while boosting the

American economy and education system. These achievements allowed NASA's leadership to produce and sustain a sociotechnical imaginary that framed human space flight as a crucial enterprise for establishing and maintaining the nation's place in the world.

Paine and the elite group of NASA managers, policy makers, and aerospace companies that shaped the space program's direction predicated this imaginary on a belief that they served citizens through their public engagement approaches. Specifically, they surmised that by communicating openly about the results and value of their efforts and meeting requests for access to information about the astronauts and space, they bred and sustained an attestive public that appreciated NASA's strides in human space flight activity. NASA officials engaged American citizens in the space program while casting them in passive roles, as consumers and supporters of human space flight activity, rather than as entities they sought to engage in dialogue about the program or who might object to NASA's carefully laid plans.

This vision of a connection among human space flight endeavors, the nation's strength, and an attestive public of American citizens helped to sustain NASA's achievements in the 1960s. But contrary to Paine's aspirations, NASA faced considerable resistance in obtaining citizen and political support for a major human space flight initiative beyond Apollo. Belying NASA's view of an attestive American public, citizen interest in human space flight had not proven universal and became increasingly fragmented as the decade wore on. Although NASA and federal policy makers had stood behind Apollo to promote American democracy in the Cold War heatup of the 1960s, a costly, elitist, and spectacle-oriented human space flight program, no matter how successful technically, ran counter to the national mood of the late 1960s and early 1970s, which emphasized pragmatism, attention to basic human needs, equal opportunities for all, and skepticism about state uses of technology. In this chapter I examine how these societal shifts affected NASA's post-Apollo human space flight program options. I explain why and how, rather than receiving the go-ahead to go to Mars, NASA secured a political commitment to just one piece of the ambitious program Paine espoused: the space shuttle.

Examining Options for the Future of Human Space Flight

Entering office six months before the Apollo 11 launch, Nixon knew he would need to decide the future direction for the civil space program. His transition task force on space had weighed in with recommendations, but Nixon's economic advisor, Arthur Burns, proposed that the president form an interagency committee to sift through the options in greater detail. In February 1969 Nixon tasked Vice President Spiro Agnew, also chair of the National Aeronautics and Space Council, with spearheading a Space Task Group to fulfill this function. The Space Task Group included NASA administrator Paine, Defense Secretary Melvin Laird (represented by Secretary of the Air Force Robert Seamans), and Nixon's science advisor, Lee DuBridge. Bureau of the Budget director Robert Mayo, Atomic Energy Commission chairman Glenn Seaborg, and Undersecretary of State U. Alexis Johnson participated as observers.

Paine initially tried to ask Nixon directly to support a robust post-Apollo human space flight program. Nixon's advisors, however, directed Paine to work within the Space Task Group. Shortly after the Apollo 11 *Eagle* touched down on the moon, Paine presented an ambitious human space flight proposal to the Space Task Group. Paine told its members that "Apollo 11 started a movement that will never end, a new outward movement in which man will go to the planets, first to explore, and then to occupy and utilize them."[3] He then handed the floor to Wernher von Braun, who explained the details of an expedition that would take a dozen astronauts to the Martian surface and back to Earth. The journey would begin in late 1981 and return to Earth orbit in the summer of 1983.

Vice President Agnew supported a human program ambitious enough to rival Apollo. At the time of the Apollo 11 launch, he publicly expressed his feeling that the United States "should articulate a simple, ambitious, optimistic goal of a manned flight to Mars by the end of the century."[4] But even working within the glow of Apollo 11's epochal achievement, the members of Nixon's Space Task Group were not universally won over by NASA's advocacy for another costly program. To some, a quest to explore and exploit the planets did not seem to be anchored to any national priority. Sensitive to growing domestic social turmoil and budget pressures,

Seamans argued that NASA might make a more politically acceptable contribution if the agency were to apply its capabilities to "solution of the problems directly affecting men here on earth."[5] Mayo opposed it based on the president's commitment to submitting a balanced budget for 1971 during a time of reduced revenues, inflation, and the financial burdens of the Vietnam conflict and the needs of social programs.

Unable to arrive at a consensus recommendation, the Space Task Group presented Nixon with a report with options for consideration in September 1969. While the group agreed the space program was a "national resource" and acknowledged that Apollo 11 marked "only a beginning to the long-term exploration and use of space by man," the disparity in the members' policy preferences was clear. The opening summary's second paragraph noted "that the space program for the future must include increased emphasis upon space applications" in areas including Earth resources and communications. The next paragraph stated: "We have also found strong and wide-spread personal identification with the manned flight program, and with the outstanding men who have participated as astronauts in this program. We have concluded that a forward-looking space program for the future for this Nation should include continuation of manned space flight activity." The report outlined five program objectives that would balance human and robotic projects: an expanded space applications program; the use of space to accomplish defense missions; a program of robotic missions to study the Earth, the moon, planets, and the universe; opportunities for international participation and cooperation; and the development of space systems, beginning with a "new space transportation capability" and space station modules, focused on the principles of commonality, reusability, and economy.[6]

Where human space flight was concerned, the report made clear the opportunities and costs. Nixon's domestic policy advisor, John Ehrlichman, had discouraged the Space Task Group from including an option as ambitious and costly as Paine's proposal, but he did not object to the group raising the possibility of a long-range goal—a human mission to Mars—and other human space flight objectives.[7] The Space Task Group listed three pathways to achieve a human Mars mission before the end of the century. The most aggressive, costly option, peaking at $8–10 billion by 1980, entailed completing a modest Earth-orbiting station by 1976, a fifty-man Earth-orbiting station and a lunar surface base in 1980, and a human flight to Mars in 1983. The least ambitious option, needing half the amount of funding, would develop the modest station by 1977 and the larger station and lunar base before the mid-1980s but defer the Mars mission to an undetermined date. All three options included two foundational elements: a reusable shuttle operating between Earth's surface and low-Earth orbit in an "airline-type mode" and chemically and nuclear-propelled systems to move humans and equipment between orbits, to lunar orbit, and to the lunar surface.[8]

Sensitive to how the president might react to the high price tags of any of these options, the Space Task Group report did not specify any time scales for accomplishing these milestones or landing humans on Mars. When Agnew met with the president to transmit the report, however, he urged Nixon to support the middle option, which called for the development of a modest space station and an Earth-to-orbit shuttle by 1977 and a human Mars mission by 1986. A few days after the meeting, Paine wrote to the president endorsing this option and expressing his hope that as the nation progressed toward meeting its other needs, "we might be able to reexamine this and move closer to Option I."[9]

With the Space Task Group's report in hand, Nixon had to determine what sort of future to champion for NASA, and to what extent human ventures off the planet should be a part of that vision. Nixon's engagement with the space program to date suggested that the president might have been willing to embrace the challenge of sending humans to Mars. As Dwight Eisenhower's vice president he had played an advisory role in establishing the civilian space program. During his 1968 presidential campaign, he had given a speech at the Manned Spacecraft Center in Houston on his determination to keep the United States as the world's leader in space, implying that human space flight would play an important part.[10]

Moreover, like many Americans and people around the world, Nixon had felt the thrill of Apollo 11's success. The president lauded the lunar landing's significance for the nation and for humanity. He talked with the Apollo 11 astronauts just prior to launch, while they were on the moon, and following splashdown. During the third exchange he claimed that

Apollo 11's journey had represented "the greatest week in the history of the world since the Creation." Seeking to celebrate the accomplishment on a countrywide scale, Nixon designated Monday, July 21—the day following the lunar landing—a National Day of Participation. He directed federal agencies to close and fly American flags and encouraged state and local governments, private employers, and school systems to do the same so that citizens could "share in the significant events" of that day. Noting that television "makes all of us participants" in space flight, Nixon's proclamation emphasized that the enterprise was commonly owned and unifying: "The adventure is not theirs alone, but everyone's; the history they are making is not only scientific history, but human history. That moment when man first sets foot on a body other than earth will stand through the centuries as one supreme in human experience, and profound in its meaning for generations to come."[11]

His outward enthusiasm notwithstanding, Nixon was an astute politician. He paid close attention to public opinion polls, the media, and the sentiments of Congress as he contemplated policy alternatives. Space was no exception, and the president was determined to forge a path that fit best within the interests of various constituencies. Nixon would spend the next several months monitoring reactions to the Space Task Group report to inform what sort of vision for space flight and resources to include for NASA in his 1971 funding request to Congress.

When establishing the Space Task Group, Nixon had told member officials that "in developing your proposed plans, you may wish to seek advice from the scientific, engineering, and industrial communities, from The Congress and the public."[12] The suggestion to include "the public" reveals the president's interest in looking beyond NASA's political and financial stakeholders. The Space Task Group chose to fulfill Nixon's recommendation by proxy, inviting a group of thirty-one business, community, and thought leaders to advise the committee on directions for the space program. Agnew told those convened that while the Space Task Group would not be using a process of "participatory democracy," he saw them as "representing the man in the street" and asked for their advice on space policy decisions that "we hope the man in the street will be happy to live with for the next decade."[13]

After receiving briefings on the Space Task Group's plans, several of the invitees questioned the necessity of creating a major new goal for NASA, doubting its potential impact on public interest in the space program. Some major industry and science stakeholders also expressed concerns. The American Institute for Aeronautics and Astronautics rejected the wisdom of pursuing a single, major space objective given that it would restrict other pursuits, including activities in Earth orbit.[14] The National Academy of Sciences' Space Science Board, meanwhile, expressed skepticism about a long-term human space flight program. The board's input to the Space Task Group acknowledged "possible roles for man" in future scientific activities but advocated in the near term for studying the solar system and universe using robotic probes and telescopes and applying space technology for "the economic and social uses of mankind."[15]

It is unclear whether these inputs ever reached Nixon; the Space Task Group report did not acknowledge them or how they were considered. What is notable, however, is that neither the space-related professional organizations nor the people intended to represent the views of "the man in the street" advocated for the spectacular human space flight program envisioned by Paine and Agnew. Rather, their perspectives reflected the concerns that some Americans held about undertaking another large-scale human space flight effort. While Apollo 11's extraordinary achievement ignited public euphoria and pride in humankind's capabilities, those sentiments did not ring true for all Americans and proved ephemeral for many. Nixon would have to determine how to relate any human space flight program to a citizenry that was *not* attestive but in fact divided concerning this initiative.

Facing Challenges to Human Space Flight's Legitimacy

To be sure, many Americans subscribed to NASA's Apollo-era imaginary and believed in maintaining a strong national space program. But polls revealed that a significant number of respondents doubted the prudence of the nation's resource commitment. Even in 1969, at the time of the lunar landing, only 53 percent of those surveyed believed the mission justified the investment.[16] Another poll showed that 39 percent of respondents favored a government-funded mission to Mars while 53 percent opposed.[17] A

July 27, 1969, *New York Times* editorial echoed these themes, stating: "This is not, in our opinion, a time to commit the nation to any hard timetable for another space spectacular, or to launch a crash program for Martian landing comparable to the convulsive effort that has put us on the moon."[18]

Indeed, questions about the space program's value relative to other national concerns became more poignant as two men bounced about on the moon's surface at a cost to taxpayers of tens of billions of dollars. Many liberal intellectuals and disadvantaged groups maintained that the nation had misplaced national priorities in supporting the space program. They asked: If we can put a man on the moon, why can't we [solve any number of problems on Earth]? Physicist Ralph Lapp observed, "Our astronauts glide through space at 25,000 m.p.h. while our streets are choked with bumper-to-bumper traffic. . . . Space men nibble on expensively developed diets while Biafran children die from malnutrition."[19] Reverend Ralph Abernathy, president of the Southern Christian Leadership Conference, made this point in a visceral way when he brought dozens of poor southern families via mules and wagons to Kennedy Space Center to protest in the days before the Apollo 11 launch. Abernathy lamented the "bizarre social values" of an American society that "has the capacity to meet extraordinary challenges," but "has failed to use its ability to rid itself of the scourges of racism, poverty, and war, all of which were brutally scarring the nation even as it mobilized for the assault on the solar system" (see fig. 2.1).[20] The antipathy many Black Americans felt toward the Apollo program while they endured life in inner city slums was further brought to light by Gil Scott-Heron's 1970 spoken-word poem "Whitey on the Moon."

Members of Congress also raised concerns about NASA's neglect of society's most dire needs. Republican senator Mark Hatfield of Oregon remarked during an August 1969 Senate Space Committee hearing that while he did not want the nation to abandon the space program, the decision to send humans to the Moon "showed a distorted sense of values in this country. I think a lot of people are thinking this every day: 'We have unmet needs elsewhere—in education, housing, and so forth.' They began to see the space program as the detractor."[21] Representative Joseph Karth of Minnesota argued that a mandate to put humans on Mars would

FIGURE 2.1. The Rev. Ralph Abernathy, Hosea Williams, and other civil rights activists hold a protest on the steps of a model of the lunar module at Kennedy Space Center on July 15, 1969, the day before the Apollo 11 launch. Photograph from Bettmann Archives, Getty.

indicate a "complete lack of respect for the taxpayer."[22] Even Senator Edward Kennedy, brother of the late President John F. Kennedy, who had spurred the Apollo program, called for a deceleration of the program once the human moon landing was achieved and to divert a "substantial portion" of NASA's budget "to the pressing issues here at home."[23] Others criticized Apollo as a playground and welfare program for the aerospace industry.

Doubts about human space flight also emerged among those who had grown skeptical about government and military uses of science and technology. Americans living in the late 1960s were enjoying some of the most remarkable advances that science and technology had to offer: vaccines that eradicated deadly diseases, oral contraceptives that opened new career opportunities to women, and conveniences such as handheld calculators and countertop microwave ovens. But the nation's citizens had also seen some of technology's most devastating effects on human life, including the destructiveness of the atomic bomb and other wartime weaponry. From the perspective of antitechnocrats, the space program was another weapon of sorts—in this case for fighting the Cold War. While iconic photographs of planet Earth suspended in the blackness of space taken by Apollo astronauts bolstered the environmental movement, the impacts of synthetic pesticides and air pollutants that needed to be mitigated were formidable. On the eve of the Apollo 11 launch, journalist Joseph Morgenstern questioned the space program's wisdom in a *Newsweek* piece. "We want desperately to believe it when our political and scientific leaders assure us with such calm rationality that they know precisely what they're doing," he stated, while also lamenting "the decades of technologists who've calmly welshed on their promises of health and happiness for all."[24] Many like Morgenstern called for democratic control of technology and investments in science to help address the growing health, environment, and energy crises.

Even some human space flight enthusiasts had become disillusioned by NASA's direction. Some experienced what Michael A. G. Michaud called "a sense of frustrated hope" because of the program's seemingly elitist nature.[25] Impressive as it was to watch television coverage of the astronauts taking humanity's first steps on another world and to see them regaled in ticker tape parades, the fact that few were chosen—and from a small pool of white, male test pilots—left many people feeling disconnected from the space program.[26] In March 1971 NASA deputy administrator George Low was told by editors at *Life* magazine, who had once thrilled at holding exclusive rights to the astronauts' personal stories, that the space travelers would no longer "sell" because people could not relate to them.[27] NASA's space flight imaginary and means of realizing it had not included a meaningful role for many people.

The imaginary had become problematic in another sense. Linking human space flight with establishing America's technological and ideological supremacy over the Soviet Union created a sense for many Americans that NASA and the human program had only this single reason for being. The Apollo 11 astronauts' arrival on the moon and safe return to Earth seemed to signal the end of the space race, with the United

States its prestigious winner. Did this mean that human space flight—and perhaps NASA—were no longer needed, and that the nation could turn its attention to diplomatic efforts aimed at peace? In the words of public opinion researcher Herbert Krugman, public support for Apollo had been "designed to self-destruct" once the program's major objective was achieved. John Noble Wilford, of the *New York Times*, reflected that NASA failed to hold public interest in human space flight because the agency and the media had conditioned people to regard the lunar landing as "the grand climax of the space program, a geopolitical horse race and extraterrestrial entertainment—not as a dramatic means to the greater end of developing a far-ranging spacefaring capability."[28]

Four months after Apollo 11, doubts about the value of space spending from across the political spectrum became evident to President Nixon. An October 1969 *Newsweek* poll of Americans with middle-class household incomes revealed that 56 percent of respondents believed that the government should be spending less money on space exploration, while only 10 percent wanted to see greater spending. Nixon's liaison to NASA, Peter Flanigan, pointed out to the president: "This represents 61% of the white population and is obviously the heart of your constituency."[29] While liberals and minorities were most vocal in their opposition to a major government-supported space program, some within Nixon's "silent majority" of middle-class Americans—those to whom the president most appealed for his political strength—were likely questioning the value of space exploration. There was clearly a chasm between people's enthusiasm for the spectacle and sense of national pride that Apollo 11 conjured—where such interest had existed—and their convictions about whether the nation should be investing resources in human space flight spectaculars. If Nixon did not moderate the space program, reducing overall federal spending as he had pledged while campaigning, voters might well take that into consideration in the 1972 election.

Under these constraints, Nixon felt that he could not accept even the least ambitious of the three options the Space Task Group put forward, let alone the human Mars initiative Agnew and Paine advocated. Following considerable back-and-forth between NASA and administration officials, Nixon proposed for 1971 a reduction of more than $400 million from NASA's 1970 budget. Those cuts would have immediate impact on the Apollo program, requiring NASA to cancel the last scheduled mission (Apollo 20) and defer missions 18 and 19 until 1974. Paine and Nixon discussed the reduction in January 1970, as final budget decisions were being announced publicly. Believing strongly in the space program's long-term future while pointing to what the polls suggested about support, the president encouraged Paine to try and engender public interest in the human space flight program in much the way NASA often had, by having "astronauts visit the smaller cities of America like Rochester or Syracuse." He instructed Paine to draft a speech for him announcing plans for NASA that avoided the appearance that he was taking money from social programs and the needs of citizens to fund "spectacular crash programs" in space. Paine assured the president that the speech would illuminate that his support of space science and technology were "of vital importance to the nation's future."[30]

Nixon released his official response to the Space Task Group recommendations in a public statement in Key Biscayne, Florida, on March 7, 1970. The carefully crafted speech reflected Nixon's determination not to terminate the human space program while at the same time defining "new goals which make sense for the seventies" and making sure that "space expenditures . . . take their proper place within a rigorous system of national priorities." He suggested that "what we do in space from here on in must become a normal and regular part of our national life" while noting that these activities "must be planned in conjunction with all of the other undertakings which are also important to us." Declaring that "space activities will be part of our lives for the rest of time," Nixon concurrently stated that the space program he would support would "put our new learning to work for the immediate benefit of all people."[31] The speech offered three general purposes to guide the space program: exploration, scientific knowledge, and practical applications of knowledge gains to benefit life on Earth. All at once, the statement seemed to promise that NASA would realize the myriad of dreams, expectations, and limitations that its disparate publics demanded of it.

In trying to make these overtures appeal to space supporters and detractors alike, Nixon's speech proved almost completely noncommittal to human space flight. The president stated that "the most important thing about man's first footsteps of [sic] the moon is what they promise for the future." But the speech was silent on details, project plans, and timelines. An advocate of cooperation with other nations in space, Nixon said that "we look forward to the day" when NASA would work with the space agencies of other nations to internationalize the US space program's astronaut crews. He noted that decisions about lunar voyages beyond the remaining Apollo missions would be based on the program's results. He asserted that the nation "will eventually send men to explore the planet Mars" as part of a "bold" program to explore the planets and universe using robotic spacecraft but gave no target dates. Nor did the speech mention plans for a major orbiting space station; the president only specified that NASA would use systems developed for the Apollo program to build an "experimental space station" to prepare humans to live and work in space for long periods of time. This project, which became known as Skylab, would fly three crews of three American astronauts each to the orbiting workshop to conduct a variety of scientific experiments over one year.[32]

Indeed, the president made no pledge to the sort of major post-Apollo human space flight initiative that Paine and Agnew had desired. Aside from the remaining Apollo missions and Skylab, the only other element of Nixon's speech that resembled any of the options the Space Task Group had offered was a declaration to examine the feasibility of reusable space shuttles to reduce the cost of transporting payloads into space. Here, the president picked up on the Space Task Group's point that "much of the negative reaction to manned space flight" by scientists and members of the public "will diminish if costs for placing and maintaining man in space are reduced and opportunities for challenging new missions with greater emphasis on science return are provided."[33] Nixon emphasized in his speech that "such a capability—designed so that it will be suitable for a wide range of scientific, defense and commercial uses—can help us realize important economies in all aspects of our space program."[34] The president requested $110 million in his 1971 budget to conduct studies of such a transportation system and the space stations the system might one day serve, but he offered no guarantee of a long-term funding commitment. Through his speech and budget plan, Nixon defined the possibilities for a new sociotechnical vision involving human space flight. Over the next two years, NASA and the administration would work to craft a new initiative and imaginary that would serve a broader range of societal and scientific purposes and resonate with more American citizens.

Moving toward a Democratic Human Space Flight Program

Even with the president's demotion of the human space flight program, Paine did all he could in the ensuing months to invigorate human space flight activity. He forged international agreements with Western European nations and Japan to help show the White House and Congress worldwide interest in participating in human flight and other space projects. He also worked closely with NASA senior leaders to consider how lunar bases, Mars missions, and space stations could fit within the agency's future through the year 2000. Nevertheless, Paine resigned from NASA in September 1970. As someone who believed deeply in humanity's future among the stars, it was difficult for him to remain the head of the agency when it seemed to him impossible "to get public support for the space program, to get public support for any program except sewers."[35] The message Nixon had sent was that with the urgency of competition with the Soviets receding, the priority for NASA, let alone human space flight, was not very high. Any hope of garnering sufficient support for a human flight initiative would depend on a change in NASA's justification for human space flight and how it framed the benefits and costs to the nation.

NASA deputy administrator Low, who became acting administrator when Paine departed, shared Paine's passion for human space flight. Nevertheless, he was prepared to move away from NASA's Apollo-era philosophy of achieving success at any price. In preparing the agency's fiscal year 1972 budget request, Low asserted that NASA needed to shift from Apollo's global focus to an inward national focus, noting that the space program had to be "useful to the people here on Earth" and that human space flight could not be exempt from that requirement.[36] In April 1971 Nixon appointed NASA's new administrator, James Fletcher, a physicist who had worked in the space industry and served most recently as president of the University of Utah. Fletcher, along with Low as deputy administrator once again, worked to make cost reduction and broader relevance central tenets of their efforts to demonstrate the feasibility and value of a reusable space transportation system and space station. But as cost and political realities set in, they would direct their labors, and their hopes, toward only one project: the space shuttle.

Fortunately, Fletcher and Low had a good foundation to leverage in their quest. George Mueller, head of NASA's Office of Manned Space Flight, had recognized in 1967 that reducing the cost of access to space would be crucial to public and political support for any post-Apollo space initiative. With budgets on the decline, NASA would need to figure out how to stretch its resources to continue to press the boundaries of human activity in space. The Saturn V rockets provided tremendous launch capability but they were also extremely costly, requiring the production of a new booster for each mission. What NASA needed, Mueller surmised, was a reusable space transportation system—one that would require a considerable upfront investment for design, development, and construction but would ultimately reduce the cost per flight, as an airliner achieved. Indeed, the idea of a spaceplane—taking off vertically like a rocket, relying on wings to glide back to Earth horizontally, and landing on a runway—had circulated for years among space aficionados.[37] In 1968, Mueller delivered a speech to the British Interplanetary Society in which he predicted that "the next major thrust in space" would be "an economical launch vehicle for shuttling between Earth and . . . orbiting space stations which will soon be operating in space." Like his visionary counterparts at NASA, Mueller viewed this new craft as an essential element in humanity's movement into space, asserting: "The Space Shuttle is another step toward our destiny, another hand-hold on our future. We will go where we choose—on our earth—throughout our solar system and through our galaxy—eventually to live on other worlds of our universe. Man will never be satisfied with less than that."[38]

Mueller was as strategic as he was pragmatic and visionary. In addition to recognizing the need to bring down the costs of space flight, he also foresaw the importance of broadening the shuttle's utility to gain political and public support for the vehicle. He urged the internal task group that was established to lay out basic shuttle requirements to think in these terms. In the summer of 1969, the task group issued a report that identified station servicing as the shuttle's primary rationale but broadened its possible contributions to include several types of missions, including deploying satellites to multiple orbits; on-orbit satellite check-out, repair, and retrieval; orbital delivery of propellant; and carrying a research module in its payload bay.[39] Mueller directed NASA's shuttle study contractors to be sure their designs could support these potential uses. A two-stage, fully reusable system, with a liquid flyback booster, emerged as NASA's favored choice.

Mueller's strategy worked for gaining a foothold in the Space Task Group deliberations: the air force's Seamans, the American Institute for Aeronautics and Astronautics, and the President's Science Advisory Council all endorsed the promise that a reusable space transport system would have for lowering space operations costs and providing versatile capabilities, even if they did not support pursuing a space station. NASA nonetheless succeeded at getting the Nixon administration to request 1971 funding for feasibility studies for *both* a shuttle and a station and was pre-pared to defend that request. Agency officials testifying before Congress did not just point to the shuttle's value to science and applications but continued to link the shuttle to the station, and to the future possibility of a human Mars mission. Representative Olin Teague, a Democrat from Texas and chair of the House Committee on Science and Astronautics' Subcommittee on Manned Space Flight, for one, welcomed the proposal. Teague proposed adding $80 million for shuttle and station studies to the House's authorization bill for NASA.

The study funding incensed members of Congress who regarded an-other large-scale technological program as anathema when pressure was on to support social programs. Karth, Teague's counterpart as the chair of the House Committee on Science and Astronautics' Subcommittee on Space Science and Applications, decried human space flight initiatives' impact on funding available for science programs supported by robotic probes. Concerned that the shuttle would lead to an eventual commitment to a human Mars mission, Karth proposed an amendment to the House bill to eliminate all funding for the shuttle and station projects. It failed on a tie vote of 53–53. The case against the shuttle and station then moved to the Senate. William Proxmire of Wisconsin and Walter Mondale of

Minnesota, who felt that NASA had not been forthright with Congress during hearings following the 1967 Apollo 1 fire, were dubious of another human space flight effort. Proxmire regarded the shuttle as a waste of tax dollars. Mondale led a charge to eliminate the shuttle and station study funding in the Senate's authorization and appropriations bills for NASA; neither proposal passed, but the appropriations vote saved the funding by only a small margin: 32–28. House and Senate conferees passed a NASA budget of $3.3 billion, including the original $110 million for shuttle and station studies. The topline funding level nonetheless required NASA to make reductions. Agency officials cancelled two more Apollo missions, 18 and 19, figuring that this measure would also eliminate the risk of losing another crew on an Apollo mission—which had nearly happened with Apollo 13—and the possibility of jeopardizing Congress's support for a post-Apollo human program.

The shuttle and station study funding were a start, but Fletcher and Low still lacked resources to begin development of either project. And Mars remained a distant dream in the present sociopolitical climate. They decided that their best strategy moving forward with human space flight was to reframe their priorities. For one, they stopped talking of aspirations to go to Mars. During NASA's fiscal year 1972 budget briefings, Low averred that NASA had "no plans at this time for manned Mars landing missions."[40] They also chose to downplay the station to at least get approval for the shuttle, which would be needed to enable the station. Dale Myers, who headed NASA's manned space flight office, explained: "The only logical thing out of those three was the shuttle. Couldn't build a space station because you couldn't . . . support it. They were canceling the Saturn V, and . . . you had to have a launch vehicle. So the shuttle was the only choice we had."[41]

NASA did not just put the space station on hold; it altogether severed the connection between the shuttle and the station in its discourse to get the shuttle funded. The shuttle would no longer be a workhorse to support a space station but instead become an end in itself, providing economic and frequent access to space for satellites that would benefit their owners, users, and society (see fig. 2.2). Far less costly and complex than lunar bases or a Mars mission, the shuttle would suit the mood of

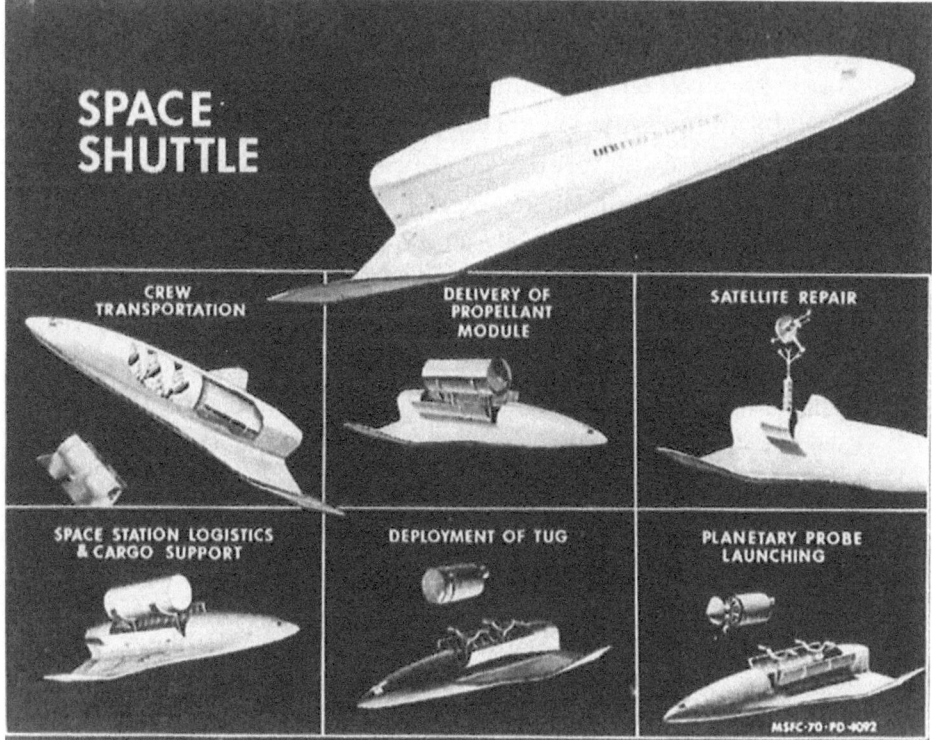

figure 2.2. A NASA Marshall Space Flight Center 1970 graphic depicts an early space shuttle design concept and the vehicle's envisioned uses. Graphic from NASA.

the times and offer something practical to the nation while the agency developed capabilities for living and working in space. NASA was on its way toward forging a new sociotechnical imaginary for human space flight based not on demonstrating geopolitical might but on facilitating the pragmatic uses of space by many.

Branding the shuttle as a versatile vehicle meant that NASA would work with customers to use the system. But there was a problem: nobody outside of NASA had expressed a serious desire to use such a vehicle. The agency would need to undertake a concerted outreach effort to drum up business from private satellite owners and researchers in industry and academia once the shuttle was approved. Getting presidential and congressional approval, however, required the agency to assure there would be enough payloads to justify building the shuttle. Fletcher, Low, and Myers therefore agreed that they needed to secure a commitment to launch military payloads—a tall order given that the Department of Defense had been content launching its payloads on expendable vehicles.[42]

The US Air Force drove a hard bargain: if they were to use the shuttle, NASA would need to ensure that the vehicle met its requirements. These included a 65,000-pound lifting capability and a 15-by-60-foot payload bay. The Air Force also demanded a 1,500-mile "cross-range" capability to allow the vehicle to land anywhere within a 3,000-mile span along its return trajectory. While the changes would make the shuttle heavier and larger and would add more than 20 percent to its development and operations costs, NASA felt it had no choice but to agree to them. Former astronaut Joe Allen called the move a "pact with the devil."[43] Following the agreement, Seamans indicated before the Senate space committee in March 1971 that the Air Force intended to recommend the shuttle's authorization and the phasing out of its use of expendable launch vehicles.

Another aspect of NASA's practical argument for the shuttle was the system's ability to launch frequently and reduce the costs of space operations, long desired by both NASA and the air force. Defenders of the president's fiscal policy, Office of Management and Budget (OMB) staff members were skeptical of the shuttle's merits, and in 1970 they directed an independent assessment of the shuttle's cost-effectiveness relative to expendable launch vehicles. Would the shuttle indeed realize cost savings, and if so, would it save enough to warrant the investment? Fletcher and his team were suddenly in the position of needing to make an economic justification for human space flight—something the agency had never needed to do during the Apollo program.

Fletcher tasked Mathematica, an economic consultancy in Princeton, New Jersey, to conduct the study. Mathematica drew on estimated costs of launch vehicle and payloads and anticipated federal and private space transportation needs projected by NASA, the Department of Defense, the Aerospace Corporation, and Lockheed.[44] In May 1971, Mathematica concluded that a fully reusable shuttle would be marginally cost-effective, breaking even by flying about six hundred missions between 1978 and 1990 and costing $12.9 billion to develop.[45] The actual savings would depend on the shuttle's flight rate, which in turn depended on the actual number of payloads demanding launch. Most savings, Mathematica estimated, would come from the shuttle's ability to reduce payload costs by accommodating payloads with simple requirements and enabling them to be repaired in orbit or returned to Earth.[46]

Mathematica's analysis did not help NASA's cause with OMB. The budgeteers were doubtful of the study's assumptions and informed NASA in May 1971 that the administration would cap the agency's annual budget at roughly $3 billion for the rest of Nixon's presidency—far from the $5–6 billion budget that NASA had enjoyed at Apollo's peak. Moreover, they specified that the shuttle's development could cost no more than $5 billion and would need to demonstrate a 10 percent return on investment. With Mathematica's cost estimate for the fully reusable shuttle's development far exceeding that amount, the agency would need to look to alternative architectures, including some expendable elements. Fletcher tasked NASA's shuttle study contractors with examining options and asked Mathematica to conduct a second study to identify factors that would make the shuttle less costly than conventional launch vehicles.

Human space flight detractors in Congress, the space science community, and the media sustained their crusades against the shuttle. Senator Proxmire believed that NASA's arguments that the shuttle would enable continuity and reduce costs of the US space program hardly justified support. He asked his fellow senators: "But why do we actually need it? What would it help us accomplish that we could not otherwise accomplish? NASA seemingly has no answers to these questions."[47] Senator Mondale pointed to an air-force-commissioned RAND report claiming that the shuttle would not prove more cost-effective than expendable launchers and would end up costing taxpayers even more. He also sympathized with the Federation of American Scientists and individual prominent scientists who had registered their opposition to the shuttle. University of Iowa physicist James Van Allen, for one, wrote to Mondale in May 1971 that shuttle advocates seemed to believe "with religious fervor" that technology should be pursued for its own sake and that whatever outcomes resulted would justify it. "Responsible public policy," Van Allen said, "requires the demonstration of specific human benefits." Van

Allen added that the shuttle was "cut from same cloth" as the recently cancelled Supersonic Transport program and ought to be subject to the same cost-benefit considerations.[48] Some journalists shared these views. Robert Cowen of the *Christian Science Monitor* called the shuttle "a first-rate concept in search of specific uses," while Englebert Kirchner penned an article for *Innovation* accusing NASA of attempting self-preservation without a clear rationale for its existence.[49]

But the shuttle also had advocates in the space science community who lauded its potential to advance their fields. Fred Whipple of the Harvard College Observatory endorsed the shuttle for its anticipated ability to launch large scientific payloads and enable humans to repair and operate major telescopes, such as the Large Space Telescope (which became the Hubble Space Telescope), then under consideration by the astronomy community. Berkeley Space Sciences Laboratory director Kinsey Anderson appreciated that the shuttle would support scientists in many fields and make the deployment of spacecraft focused on scientific research more economical.[50]

Meanwhile, several members of Congress communicated their support for the shuttle. Nebraska senator Carl T. Curtis issued a statement titled "Space Shuttle—The Key to Our Future" in which he lauded the shuttle's promise "to open the space frontier in the way the railroads opened the west, or in the way the DC 3 airplane began the great expansion of commercial aviation." The vehicle, he added, would provide "a new capability that is needed" for national security, science, and practical applications and "give new opportunities to use space to serve mankind." Taking on those who preferred to see NASA funding diverted to address social needs, Senator Howard W. Cannon authored an article in the Aerospace Industries Association's newsletter arguing that investments in space technology like the shuttle "create the strong economic base that will be necessary if we hope to solve *any* of our problems in the future."[51] Even Representative Karth backed down from his antishuttle stance. Though he still wanted to see NASA make a more concerted effort to address societal problems, he indicated a willingness to back the program based on NASA's commitment to separating the shuttle from the station project and finding a cost-effective design. This support trumped the opposition. An

amendment offered by Mondale, Proxmire, and other senators to eliminate the shuttle in NASA's 1972 authorization bill was defeated 64–22, and
Congress approved $118.5 million for continuing shuttle studies in 1972.[52]

Attitudes in the Nixon administration were also turning in favor of
the shuttle. OMB deputy director Caspar "Cap" Weinberger worried that
his staff's proposed budget cuts for the space agency were too extreme.
Weinberger wrote in an August 1971 memo to Nixon that these cuts would
suggest to the world that America was "turning inward" and "voluntarily
starting to give up our super-power status." He believed that the country
"should be able to afford something besides increased welfare, programs
to repair our cities, or Appalachian relief or the like" and argued that
programs including the shuttle ought to be funded. Still a space supporter
at the core, Nixon endorsed the sentiment in a handwritten note on the
memo stating, "I agree with Cap."[53]

Throughout the fall, as the Nixon administration formulated its 1973
budget request, justification for the shuttle grew. In October, Mathematica released the preliminary results of its follow-on study to identify a
cost-saving shuttle configuration. The analysis concluded that a particular concept, a Thrust Assisted Orbiter Space Shuttle System (TAOS),
would be economically feasible and yield the greatest savings. The system
would include a reusable space plane, or orbiter; an expendable external fuel tank; and two expendable booster rockets mounted on opposite
sides of the tank. Mathematica acknowledged that while NASA and the
Department of Defense would constitute much of the total demand, "It is
our strong belief that the major portion of space transportation demand in the
1980's will come from economic applications of space technology to meet the
growing needs of the US and other developed and developing countries." Such
needs could include scientific research, communications, Earth observations, navigation, and applications such as space-based manufacturing
processes and energy generation and transmission.[54]

Fletcher, meanwhile, continued to appeal to Nixon on grounds beyond
the economic justification. In a November 1971 paper, NASA argued five
points for proceeding with the shuttle that the agency believed would
resonate with Nixon. Aware of the president's respect for US space leadership accrued through Apollo, NASA contended first that the nation could

not forgo human space flight, claiming a "responsibility—to itself and to the free world" to continue this enterprise. Second, the paper pointed out that the shuttle was the "only meaningful" new human flight program that NASA could accomplish at current budget levels. Third, NASA maintained that the shuttle was "essential" for future programs. NASA argued that the shuttle would satisfy the range of applications the Mathematica memo stressed and suggested that it could eventually provide crew transportation and logistical support for more visionary projects, including a station in Earth orbit and orbital assembly of systems needed for lunar bases. Fourth, NASA indicated that it had done due diligence and met the Nixon administration's challenge by finding a shuttle solution costing half that of a fully reusable system. Finally, NASA pointed out that shuttle development would help the aerospace industry. Some 270,000 workers had been laid off since the peak of the Apollo program, and the shuttle would help to mitigate the impact somewhat by supporting the direct employment of nearly 33,000 people by the end of 1973.[55] With the aerospace industry employing thousands of people in key states for Nixon, including his home state of California, announcing the approval of a shuttle program could boost his bid for reelection in 1972.

The arguments were sufficient to win Nixon's support for the shuttle as NASA's next major human space flight initiative. On January 5, 1972, Fletcher and Low met with Nixon in San Clemente, California, after receiving word from the White House that the president had approved development of a partially reusable shuttle (see fig. 2.3). During the thirty-five-minute meeting, the NASA officials informed Nixon that the agency could develop a shuttle costing $5.15 billion, with a $1 billion and eighteen-month contingency, by September 1979. Nixon conveyed to Fletcher and Low multiple interests in the shuttle. The president was impressed by the shuttle's potential to fly routinely and on short notice. He was intrigued by its projected ability to "open up entirely new fields" and support civilian applications, perhaps including natural disaster response, solar power collection and distribution from orbit, and even nuclear waste disposal. The president also "liked the fact that ordinary people would be able to fly in the shuttle, and that the only requirement for a flight would be that there is a mission to be performed." He expressed

FIGURE 2.3. NASA administrator James Fletcher (*left*) shows President Richard Nixon a model of the space shuttle during their meeting on January 5, 1972, the day on which Nixon announced approval for the new space transportation system. Photograph from NASA.

interest in "meaningful participation" in the shuttle program not just by American citizens but also by foreign nations via flights of international astronauts, experiments, and hardware. At the same time, Nixon hinted at his commitment to maintaining national prestige through the program, indicating that "even if [the Shuttle] were not a good investment, we would have to do it anyway, because space flight is here to stay. Men are flying in space now and will continue to fly in space, and we'd best be part of it."[56]

That same day, Nixon issued a statement announcing that the United States would pursue the space shuttle as "the right next step for America to take, in moving out from our present beach head in the sky to achieve a real working presence in space."[57] While Nixon's March 1970 speech about

NASA's future had been light on program specifics, the new statement made clear the president's choice of direction for human space flight. But as he had done when he spoke nearly two years earlier, he cast his decision in a way designed to appeal to everyone and alienate no one. Nixon, as much as NASA, recognized the need to establish a new imaginary for human space flight and the importance of focusing not just on the new vehicle but on how various groups, including those the agency had not actively engaged in the Apollo program, would aid in legitimizing it and making it viable.

The president proffered a vision of the shuttle that tied the dreams of space flight aficionados together with the expectations of those who felt human space flight to be at best esoteric and at worst a waste: he claimed that the shuttle would "help transform the space frontier of the 1970s into familiar territory, easily accessible for human endeavor in the 1980s and '90s." The shuttle as Nixon constructed it would be relevant to everyone, "delivering the rich benefits of practical space utilization and the valuable spinoffs from space efforts into the daily lives of Americans and all people."[58] The very name "space shuttle," which Nixon preferred over the mythology-derived names of NASA's previous human programs, conveyed the president's conviction that the new vehicle would exist not for the purpose of creating space spectaculars but because it had an important job to do in providing people, experiments, and hardware with access to Earth orbit.[59]

As Nixon characterized it, the shuttle would be the people's spaceship. It represented a completely different approach to human space flight from Apollo, based on a new sociotechnical imaginary. That vision centered on the idea that the shuttle would democratize space flight, making it more accessible and relevant to more people by providing tangible benefits and enabling wider participation in the space program. Indeed, the new program would support the aims and interests of diverse groups, as it would "give more people more access to the liberating perspectives of space, even as it extends our ability to cope with physical challenges of Earth and broadens our opportunities for international cooperation." Nixon suggested that the shuttle would open the possibility that many more sorts of people could become astronauts and abandon Apollo's homogeneity: "The resulting changes in modes of flight and re-entry will make the ride safer, and less demanding for the passengers, so that men *and women* with work to do in space can 'commute' aloft, without having to spend years in training for the skills and rigors of old-style space flight." Those who questioned why NASA sent humans into space rather than relying solely on robotic vehicles could take comfort in the fact that the shuttle's capabilities to deploy, repair, and retrieve satellites would mean that the "limiting boundaries between our manned and unmanned space programs will disappear." And the aerospace industry could rest easy because the effort would "engage the best efforts of thousands of highly skilled workers and hundreds of contractor firms over the next several years."[60]

Despite Nixon's words, the shuttle continued to receive its share of criticism from those who doubted its value as the decision to approve the president's proposal for $228 million to begin shuttle development moved to Capitol Hill. Representative Bella Abzug of New York decried the shuttle's impact on social needs. Brian O'Leary, a planetary scientist who had been accepted by NASA as an astronaut in 1967 and then resigned in 1968 because he was dismayed by his chances of ever flying, raised questions about the shuttle's goals, seeing them as unsteady, unclear, and out of touch with human values. Journalist and pundit Daniel Greenberg contended that the shuttle "can be likened to a gold-plated limousine to deliver small bundles: once built, its existence becomes the justification for delivering lots of bundles." Bob Cromie of the *Chicago Tribune* argued against investing billions in a spaceship that would soon enough become obsolete.[61]

Strong advocacy for the shuttle, however, persisted. Like Nixon, Senator Henry Jackson of Washington lauded the shuttle "as a good investment which will return its costs to the people many times over." The AFL-CIO latched onto the employment benefits NASA forecasted and urged Congress to fund the shuttle on the premise that it would provide tens of thousands of American jobs.[62] Newspaper editors from regions of the country likely to benefit economically from the shuttle expressed strong support. The editorial staff at the *Miami Herald* lauded the shuttle's promise to give Brevard County, home of Kennedy Space Center, "a welcome lift out of the doldrums" brought on by recent cutbacks in space spending. The *Times-Picayune* cheered the fact that the program would reactivate the Michoud Assembly Facility for rocket construction near New Orleans.[63] Even without a local space industry presence, some city newspapers opined on the shuttle's value for the nation scientifically, economically, and geopolitically.[64]

Mathematica's final assessment was released a few months after Nixon's announcement. It provided welcome news to shuttle developers and proponents that the vehicle would be economically justified in the TAOS configuration if it flew between 300 and 360 times from 1979 to 1990, or about twenty-five to thirty flights per year. The consultants maintained that with NASA and the Department of Defense projecting more than six hundred payloads during this timeframe, the shuttle would save the nation an average of $13.9 billion in 1970 dollars. Even reducing the forecast by at least one hundred payloads, they said, would yield a savings of approximately $10 billion.[65]

Mathematica did not disclose that the payload demand figures its analysts had used were best-case scenarios: they applied only if NASA and the Department of Defense were able to get funding for their own payloads and could develop them on the timetables they outlined. The consultants also did not consider that the Manned Spacecraft Center's shuttle planners did not believe that the vehicle's turnaround requirements would realistically allow for anywhere close to sixty flights per year.[66] Nevertheless, the arguments for continuing the nation's human space flight program prevailed, and Congress ultimately came down in favor of initiating development of the shuttle.

Although NASA's human space flight visionaries would not lose sight of their long-range aspirations, they tucked away their blueprints for a station and more for the time being. Instead, they turned their attention to developing the shuttle with a new sociotechnical vision that would define how NASA interacted with the public. Legitimizing human space flight and making it viable economically and politically henceforth rested on NASA's commitment to considering citizens' diverse needs, values, and potential contributions to the shuttle program. The agency prepared to engage more people with human space flight, and in a wider variety of ways, than it had during the Apollo program.

3

Sharing the Shuttle through Discourse

In 1971 University of Alabama space policy researcher Charles
Lamb advised NASA not to worry about public opinion after he found
little correlation between polling data and the agency's budget up through
the early 1970s. Instead, he maintained, NASA should "push more vig-
orously for patrons from within the budgetary process." Lamb added
that state authorities are often slow to change direction on major program commitments, particularly when powerful players put pressure
on government decision-makers to maintain funding.[1] Indeed, both the
Apollo program and the decision to continue human space flight through
the space shuttle program were largely advocated by government and
aerospace industry leaders without strong support from most American
citizens. The fact that more than fifty thousand NASA and aerospace industry personnel in almost every state of the nation contributed to the
shuttle's development by 1977 kept up the program's political inertia, even
when technical challenges affected cost and schedule.[2]

What Lamb overlooked, however, is that national decision makers
still cared about enrolling Americans in the human space flight effort as
attestive supporters. Recognition of vocal naysayers across the country

in part led President Richard Nixon and Congress, at NASA's urging, to select the shuttle, pitched as a broadly capable vehicle, as the next step for human space flight. Although the space program hardly ranked as a major policy issue, it consumed considerable federal funding while its societal benefits remained unclear to many. Politicians who wanted to demonstrate fiscal responsibility—and win reelection—could not disregard the public backlash to Apollo and the low priority that polling numbers accorded to the space program amid other national needs. Representative Bob Casey of Texas, whose district housed the Manned Spacecraft Center, noted in House floor debate on NASA's 1972 budget that, at least to some extent, "nationwide, man-in-the-street type support" and public enthusiasm about issues including the space program influence Congress's funding decisions.[3] Casey's comment echoed what James Webb had said of George Mahon wanting to know that he had the backing of his constituents in Lubbock.

Consequently, after the shuttle program gained approval, NASA and aerospace industry officials continued to cling to the idea that a causal link existed between public support and political sustainment of space activities. Accordingly, they continued to focus on creating positive public sentiment. Grumman Aerospace Corporation, one of NASA's shuttle study contractors, asserted in a 1971 brochure: "To be truly effective, a continuing long-term space program needs the broad based support of the American people and their representatives in Congress." NASA public affairs chief Brian Duff agreed, maintaining that broad public support made possible the things NASA wanted to do. Similarly, NASA administrator James Beggs, who served during the first few years of shuttle operations, told the *Washington Times* in 1983: "We're in the business of producing shuttles, but if we are to keep producing them, we need the support of the American public." Perhaps John F. Murphy, NASA assistant administrator for legislative affairs, put it best when he unabashedly responded to J. Scott Brownell of Branford, Connecticut, the same year, stating: "It is through enthusiastic supporters of the space program, like you, that NASA makes new friends."[4]

But how would NASA go about building connections with the American people for the shuttle program and ultimately legitimize this and follow-on human space flight initiatives? As was the case in the 1960s, NASA officials and members of Congress still largely believed that communicating the agency's accomplishments, goals, and plans for the shuttle would serve those who were already enthusiastic and, perhaps more importantly, enlighten and thereby change the opinions of the many skeptics throughout the country. By 1970, half of Americans believed that the nation was spending too much on space exploration, and that fraction grew as the decade wore on.[5] NASA administrator Thomas O. Paine stated in seeking Senate support for expanding the Kennedy Space Center visitor center in 1970: "We believe that continued and improved exposure to the space program will provide a solid basis for stimulating an improved understanding on the part of the public of our program and the benefits to be derived from them." The House Science and Astronautics Committee and its Subcommittee on NASA Oversight included an additional $4 million for the agency's public affairs office in its 1972 authorization, feeling that citizens often showed "a lack of understanding" about what it took for NASA to be successful.[6]

Polling experts continued to suggest throughout the decade that sharing information was key to winning over the public. Northern Illinois University political scientist Jon Miller, who studied public attitudes toward science and technology, posited that NASA had the potential to expand the ranks of space program supporters—and hence political backing for NASA's programs—by gearing information to satisfy those less knowledgeable about its activities. Public opinion pollsters told administrator Beggs that the way to get past the fact that public support for NASA "is about a mile wide and an inch deep" was "to get more stuff out there."[7]

While space enthusiasts continued to write to NASA and express their excitement about space flight, the agency's attempts to garner attention to the few human missions taking place in the mid-1970s—Skylab and the 1975 Apollo-Soyuz cooperative mission with the Soviet Union that served as an element of Nixon's peace plan—fell flat. Television networks broke with tradition and did not broadcast the splashdown return of the third and final crew of Skylab astronauts. NASA tried to play up Apollo-Soyuz by issuing press releases, conducting status briefings throughout the mission, and sponsoring a movie about it narrated by actor Yul Brynner, but the initiative failed to grab many headlines.[8] Throughout the 1970s, NASA officials persisted in contemplating how a steady stream of news summaries, television and radio exposure, movies for commercial airliner flights, and other means of publicity could help boost general interest in the space program.[9]

One thing NASA ultimately recognized was that a blanket approach to sharing the shuttle with Americans would not be effective. As the House Science and Astronautics Committee noted in April 1971, "geographical, societal and occupational divisions of public interest create demands for varying types of information."[10] Indeed, public reactions to the Apollo program had made clear that a singular attestive American public did not exist. High-level NASA leadership were advised by their staff members in the mid-1970s that the agency needed to target public communications to resonate with "the basic value systems" of various societal groups.[11] The agency needed to focus not just on the medium but on the messages— namely, how the shuttle would live up to its imaginary of a vehicle that would capably serve many needs and interests.

It would take several years for the shuttle to begin missions and be able to deliver on that vision. As I show in this chapter, NASA employed multiple messages during the shuttle's development stage and early years of flight operations to communicate to a broad range of publics that the vehicle would be relevant and accessible to them. Its approach involved articulating anticipated benefits and openly communicating achievements—tactics with origins in the Apollo era but that NASA substantially refined for use during the shuttle program. The agency centered its discourse on four major themes to impress upon Americans the significance, value, and legitimacy of the new human space flight program. Officials reached out to specific groups using messages about the shuttle that would resonate with them, emphasizing various facets of the democratic imaginary for the shuttle in an effort to enroll them as constituencies.

Promising Practical Benefits

The public and political differences surrounding the future of government-sponsored human space flight had made clear to NASA, Nixon, and members of Congress that NASA would need to need to show the public how its programs, including the shuttle, were relevant in a social milieu that had little tolerance for extravagance, elitism, and esoteric projects. Given so many problems on Earth, NASA's focus on sending humans into space seemed insular and out of touch to many. These public perceptions were corroborated by studies that NASA commissioned in the early 1970s to better understand political attitudes and social values among the American public. A 1971 study by the Hudson Institute found that the agency was less likely to secure federal funding because its work did not directly benefit people's lives and because minority groups, youth, and a new left counterculture opposed the military-industry complex supporting it. A follow-on study concluded that while Americans saw scientific progress and industrial capabilities as instrumental to national economic well-being and peace, they did not associate NASA and the space program with their own problems or goals.[12]

These findings indicated to NASA leadership that rational explanations and social context were the order of the day. Paine conceded in 1970 that "our spectacular achievements in space have overshadowed the less dramatic but equally important story of the many benefits the nation is realizing from the space program."[13] In following Paine, administrator James Fletcher declared in 1973 that it was time for NASA "to move back from the spectacular" and "become more like one of the service agencies of government."[14] Apollo astronaut Harrison Schmitt, meanwhile, argued that NASA public relations efforts had focused too much on events, spectacles, and astronauts at the expense of providing an understanding of NASA's broad societal impact and significance for the future.[15] George Low, then serving as Fletcher's deputy, recognized that philosophical arguments about the human desire to explore "may resonate with certain social circles" but that the "average American" was "more interested in what this means to them today than in broad promises of the fruits of future explorations."[16]

NASA's public affairs chief John Donnelly agreed. He believed the agency had overemphasized discrete projects and hardware while not communicating their purposes. Donnelly recommended to Fletcher in 1973 that NASA move away from communicating in "project-oriented

mode" and employ a unifying rationale for space flight that would resonate with the "non-cognoscenti."[17] A few years into shuttle development, Donnelly expressed his belief that NASA could not simply focus
on "this new exciting piece of machinery" but instead needed to articulate its value proposition. NASA had to be able to answer the question,
"What good is it?" to sustain public support, especially after the first
few flights.[18]

Perhaps the House Committee on Science and Astronautics articulated the concern best. According to its report authorizing NASA's 1972
appropriations, "NASA has done a good job in bringing to the public the
'what' of the space program, but has not been effective in explaining the
'why' of the space program." The report noted that "the American public
will support the space program, but only if the true story of space and its
related benefits is more effectively brought home."[19] In 1974 the committee
issued another report stating that "NASA should be doing much more
in the area of disseminating space benefits information to the public."[20]
Like Donnelly, the committee wanted NASA to focus less on reporting
program status and more on explaining the benefits of its activities and
how its technologies were serving societal needs. During this time, the
House committee conducted hearings reviewing tangible benefits from
the space program and published the proceedings in reports they distributed in response to citizen inquiries.[21]

The tone and substance of communications coming out of NASA
changed drastically to reflect the new focus on pragmatic benefits of
the shuttle and other programs. Fletcher conveyed the theme in a 1972
statement: "We now look ahead to several decades of a highly rational
use of space. The focus will be on domestic needs, and the turning of our
rapidly developing space capabilities to useful work. We have made our
new program relevant to the needs of modern America." A few years later,
Fletcher told the *Salt Lake Tribune* that "now" was shaping NASA's direction. "Congress, reflecting the wish of the people," said Fletcher, "wants
it that way and that became the direction we (NASA) took." NASA Headquarters manager Jesco von Puttkamer acknowledged that the shuttle
and its focus on commercial uses of space came "in direct response to the
down-to-earth needs and demands of people everywhere."[22]

Down-to-earth needs permeated NASA communications. Speeches and other public relations materials highlighted NASA's role in developing technologies that led to useful, everyday applications, or "spinoffs." The agency started producing an annual Technology Utilization Program report in 1973, which morphed into the brochure *Spinoff* in 1976 and was made available to citizens who asked about NASA's societal benefits. Agency news releases featured technologies derived from the Apollo program such as lightweight breathing gear for firefighters, astronaut food delivery systems that could be adapted to assist the elderly, and emergency diagnosis and treatment systems.[23] Even NASA's internal newsletters echoed the emphasis on public value. *NASA Activities* introduced a section called "Are you getting your 1 cent worth from NASA?"—a reference to the percentage of each federal dollar budgeted for NASA. Johnson Space Center's *Roundup* newsletter advised employees on how to answer public questions about NASA's societal benefits.[24]

Nixon had urged Fletcher and Low to emphasize the shuttle's contributions to civilian applications. Accordingly, whereas publications of the 1960s focused on the agency's project plans and hardware without much explanation of their rationales, NASA and its contractors couched the shuttle in terms of pragmatic societal benefits. A 1972 NASA brochure about the shuttle made plain that "after a decade highlighted by driving effort and dramatic achievement, America's space program is shifting emphasis. Now the goal is practical benefits for people on Earth." It communicated that the shuttle, as a reusable system with a large carrying capacity, would make carrying, deploying, and retrieving satellites in orbit simple and economical. A later publication emphasized the shuttle's ability to be "sent off quickly on a special mission to gather information needed in an emergency on Earth, such as a flood or crop blight." A 1976 brochure from Rockwell, the shuttle orbiter's prime contractor, explained that satellites and laboratories carried into orbit by the shuttle could support forest and cropland management, discovery of new energy resources, communications, and natural hazards detection and warning. Grumman pointed to the shuttle's value for deploying military peacekeeping payloads, cooperating with other nations, and assisting developing countries in broadening their communications networks and services.[25]

SPACE SHUTTLE

For Down To Earth Benefits

◀ ▲ FIGURE 3.1. Brochures such as Rockwell's 1974 *Space Shuttle: For Down to Earth Benefits* illustrate the many societal applications NASA and its contractors envisioned for the shuttle. Reproduced by permission of Boeing Company.

NASA made a point to link the shuttle to the era's top issues, including environmental health and energy. A 1975 NASA publication indicated: "When the Space Shuttle becomes operational in 1980, it will be an important tool to provide mankind with information to help in managing and preserving our crowded Earth. . . . Payloads launched by the Space Shuttle will provide practical data that will affect both the daily lives of people and the long-term future of mankind." Rockwell joined NASA in promoting this theme. In 1974 the company published a brochure titled *Space Shuttle: For Down to Earth Benefits*, which noted space exploration's importance to environmental and natural resources management (see fig. 3.1). Rockwell later made an even more urgent link between the shuttle and stewardship of the Earth, stating: "The quality of life here—indeed, our very survival—depends on preservation of [the Earth] system. And to preserve it, we must find ways to use it more wisely. In this context, the Space Shuttle may be the most important spacecraft ever developed."[26]

Shuttle proponents also emphasized the economics of the vehicle. A 1972 NASA fact sheet estimated development, production, and facilities costs required through 1990 at $8.3 billion. Assuming the shuttle would fly some five hundred missions during that period using five orbiters that would each be used one hundred times, NASA concluded that the vehicle would save $12.4 billion over use of existing Delta, Titan, and Atlas-Centaur expendable launchers and would more than pay for itself. The net result, according to NASA, would lower launch expenses to $160 per pound to orbit—nearly an order of magnitude below current costs. Grumman's 1971 shuttle brochure noted that NASA's fiscal year 1971 budget of $3.4 billion—less than half of which would support shuttle development—represented a small fraction of what the nation would spend on social programs that year.[27] Arizona senator Barry Goldwater remarked at the unveiling of the first orbiter in 1976 that "the Space Shuttle is going to be a better deal for America than the purchase of Alaska, which was a very good bargain."[28]

The shuttle investment would have a sizable impact on the nation's economy, NASA and its contractors contended. Citing one of the many economic impact studies the agency supported, Fletcher conveyed to Congress that "for every $1 that you invest in a high technology program like the Shuttle, or like Apollo, you receive $7 back."[29] And while the shuttle would have an unearthly destination, the funds spent building it "would remain on our home planet," McDonnell Douglas's July 1979 ad in *Forbes* reminded viewers. A Manned Spacecraft Center form letter to respond to public inquiries stated, "What we are really buying is the time and talents of thousands of Americans who work on the space program." Meanwhile, Rockwell's 1976 shuttle brochure, the cover of which boasted an orbiter deploying a satellite against a background of waving Stars and Stripes, pointed out that the shuttle was being constructed by companies in almost every US state. Grumman estimated in 1971 that the shuttle would boost the purchasing power of middle-class aerospace workers and have an economic impact of $37 billion over fifteen years.[30]

NASA frequently alleged that the investment in the shuttle would be more than justified by the resulting benefits, even if it was not knowable at the time. A 1972 publication claimed: "The Shuttle represents an investment in mankind's future. It can provide dividends that will continue for decades to come." Similarly, NASA reported to Congress in 1973: "We can say with confidence that the research and development expenditures for the Shuttle will provide a considerable number of new procedures and products that will benefit all of our people in the coming decades—and all as bonuses of a transportation system that is planned to save us money directly in its application to space." NASA also contended that "scientific leaders believe the most significant benefits to people on Earth will come from inventions not yet conceived, which will be stimulated when the Shuttle makes space flight simple, less time consuming and less expensive." Rockwell joined in attesting to the shuttle's unspecified potential. The company's president and chief executive officer Robert Anderson made an analogy between the shuttle and the automobile, whose inventors could not "imagine all the wonders" it produced. Anderson also predicted that the shuttle would become "as vital to the nation's future in space" as ships, trucks, and airliners were to Americans' standard of living.[31]

Agency officials also spoke to members of groups they believed would be influential in spreading word about the value of the shuttle and the space program. In 1974 Fletcher and Low initiated meetings with industry and community leaders around the nation to communicate "the current and potential benefits of the space program on society and the nation's economy." Through these dialogues, the NASA leaders strived to persuade prominent individuals to appreciate NASA's value. They worked with the most supportive participants to host meetings in their communities to share NASA's messages and assist the agency in combating "the problem of gaining public support for space." NASA officials also embraced the opportunity to meet with student groups. Low recognized college students as a group that was to be "highly skeptical about the value of technological enterprises in general and the space program in particular" and aimed to engage and equip students interested in science and technology to speak well of the shuttle and the space program to their peers.[32]

Fletcher also saw fit to help establish a group of space boosters dedicated to attaining broad public acceptance of the benefits of the shuttle and other NASA programs. Fletcher and Edward Z. Gray, NASA's assistant administrator for industry affairs and technology utilization, worked with the National Space Club, founded within the aerospace community to promote excellence in the space program, to initiate such an association. Gray maintained that the new organization would benefit from "substantial but low profile" aerospace industry participation yet needed to have a "strong grass roots character which would draw on a broad spectrum of society" for its leadership and membership.[33] Wernher von Braun, who had since left NASA, became the first president of the new association, initially incorporated as the National Space Association and renamed the National Space Institute (NSI) in April 1975. Careful to avoid appearing to endorse the organization, Fletcher sent a letter to von Braun welcoming NSI's assistance in helping "the American public to know and understand the value and benefits" of the space program and indicating NASA's willingness to cooperate with NSI "wherever feasible."[34]

NSI's first chairman, ABC news anchor Hugh Downs, noted at a September 1975 Senate hearing that NSI filled a "communications void" between NASA and the public, including doctors, farmers, environmentalists, and others. According to an NSI pamphlet, the organization would tell Americans about the "practical uses of space by people on Earth" and how the space program would help "to solve the problems of energy, inflation, and others." Von Braun told attendees at NSI's first annual meeting that the organization would serve as "a catalyst between the space technologist and the user," making citizens aware of new space applications. Concomitantly, he said, as an association financed not just by special interests but also by individual citizen members, NSI would represent all of its supporters' sentiments when testifying before or meeting with members of Congress.[35]

Assuring an Exciting, Collective Future in Space

Not all Americans had to be convinced of human space flight's value. The space program had, beginning in the Mercury days, ignited the imaginations of citizens who followed NASA's astronauts and their missions with unabated excitement. Many had become armchair participants in the adventure through the agency's open information program. NASA's achievements had also propelled the dream of space travel into American popular culture, keeping the public engrossed through television shows such as *Star Trek*, *I Dream of Jeannie*, and *The Jetsons*. TWA and Eastern Air Lines advertised trips to the moon in the not-too-distant future, and tens of thousands who hankered for the real thing put their names on waiting lists. Many space enthusiasts wrote to NASA asking how they could become astronauts. Beginning in the early 1970s, some channeled their interests by joining with likeminded people in NSI and grassroots organizations dedicated to expanding humanity's future in space.

Scientists and engineers added to the excitement by raising the prospect of space colonization. German-born rocket engineer Krafft Ehricke's "extraterrestrial imperative" maintained that it was human destiny to expand into and exploit the solar system, while Princeton physicist Gerard O'Neill advanced designs for space colonies that could create Earthlike environments for inhabitants.[36] Members of space advocacy groups such as the L5 Society—named for a gravitationally stable point along the moon's orbit at which the society's founders envisioned placing an O'Neillian space colony—believed they would get the chance to live in space in their lifetimes. Science fiction aficionados and *Star Trek* fans envisioned similar futures.

Some space supporters expressed concern that NASA's shuttle imaginary, which conceptualized the vehicle as capable of making space flight routine and ordinary, could have the collateral effect of creating a dull space program. A *Boston Globe* guest writer maintained: "The Space Shuttle will definitely not fill our hunger. In it, astronauts will just become another group of blue-collar workers" doing "tedious" work.[37] NASA officials thus strived to depict the shuttle to space enthusiasts as a keystone to futuristic developments. Although NASA had throttled back on linking the shuttle to a space station and almost never talked publicly anymore about sending humans to Mars or other planets, Hans Mark, then NASA Ames Research Center director, suggested to Fletcher in 1976 that the agency harness burgeoning interest in science fiction to develop "new political support." Believing that science fiction had helped to sell Apollo,

Fletcher agreed it was worth discussing.[38] A few months later, NASA teamed with O'Neill to conduct the first of a series of studies at Ames on space settlements and industrialization. The participants considered how the shuttle could be used to assemble colonies and transport inhabitants and how refurbished shuttle external fuel tanks could house workers in an orbiting manufacturing facility. NASA soon was communicating these possibilities publicly, noting that the new vehicle could deliver modular units for "self-sustaining settlements" used to operate solar power stations or for manufacturing drugs, metals, or other products.[39] Rockwell suggested that seemingly fantastical uses of the shuttle such as ferrying hardware to enable processing facilities, hospitals, and climate-controlling applications in space would begin "in our own lifetimes."[40]

NASA officials made a point to reach out to space enthusiasts to explain that the shuttle would be useful in these pursuits. One community of interest was the legions of *Star Trek* fans, many of whom were idealistic youths who convened at dedicated conferences across the country. Von Puttkamer, who spoke at multiple *Star Trek* conventions throughout the 1970s, explained that some at NASA initially "didn't see the connection between us and *Star Trek*." Von Puttkamer nonetheless told his colleagues that *Star Trek* could help communicate what the space program was all about, noting that Trekkies were "not just living in a fantasy world but were very inquisitive about the country's capabilities and future in space." He told attendees at a 1976 convention in Washington, DC, that NASA wanted to meet "humanistic as well as scientific needs" and "include a little of the dreams of young people" in the space program.[41]

Likewise, John Yardley, NASA's associate administrator for space flight, stated at a 1976 NSI meeting that it would be a "cinch" to establish colonies on the moon or at L5 "if there was appropriate public support."[42] Recognizing that space settlements also could appeal to environmentalists, utopians, pacifists, and others as ecological and apolitical havens, Fletcher publicly asserted that space colonies would possibly emerge by 1980. However, he tempered his statements with the air of public responsibility that pragmatists demanded, conceding that NASA had no current plans to establish space colonies because "people are not ready to bear the costs."[43]

Indeed, NASA sometimes struggled with how best and how much to connect the shuttle to a democratized space future. For instance, NASA associate administrator Homer Newell was pleased to provide the Committee for the Future (CFF), a grassroots group dedicated to seeing humanity branch out from Earth, with outreach materials when they expressed a desire to promote public awareness of space starting in the late 1960s. Then, in 1971, CFF proposed to generate public support for the shuttle and initiatives to take humans further into space by organizing a "citizens'" mission to the moon. CFF envisioned that NASA would give them unused Apollo hardware; CFF would then conduct its mission and sell room on the spacecraft for scientific experiments, as well as lunar material and photographs resulting from the mission, to fund operations.[44]

Although several senior officials at NASA Headquarters and the Manned Spacecraft Center were fascinated by this strategy for creating a constituency, NASA acting administrator Low rejected it, citing cost, safety, and logistical concerns. Dale Myers, associate administrator for manned space flight, suggested that the group instead participate as scientific users of the shuttle when available. When CFF later contacted NASA about holding a televised conference on humanity's future in space at Kennedy Space Center near the Vehicle Assembly Building, where the shuttle would be integrated, NASA deputy associate administrator John Naugle advised the group that conference discussions should avoid suggesting that NASA was planning to colonize space or send humans to the moon. Instead, Naugle said, they should focus on space activities' roles in solving global problems.[45] Clearly, the agency still tilted to the pragmatic and was hesitant to accept the assistance of even its most ardent supporters when it could not fully control messages.

Sometimes NASA's hype of the shuttle led space and science fiction enthusiasts to channel their zeal into activist efforts that went against the agency's preferences. Months before the first shuttle orbiter's 1976 debut, some *Star Trek* fans petitioned to name the vehicle *Enterprise* in homage to the television show's famed starship. NASA had been planning to name the orbiter *Constitution* in honor of the nation's bicentennial. After tens of thousands of signatures poured into the White House advocating the name change, President Gerald Ford informed Fletcher that he preferred

Enterprise.[46] Obligated to honor the president's desire, Fletcher renamed the new orbiter accordingly. Agency officials felt mixed about the name. Some expressed concern about its implications. As one NASA official put it: "Here come the exploiters." Others, some of whom were *Star Trek* fans or at least were understanding of the attention the new name would bring to NASA, embraced its ability to popularize the shuttle. NASA shuttle program director Myron Malkin told the *Washington Post*: "This way we get a ready-made public."[47]

But the largest issue NASA officials would confront concerning the involvement of broader groups of people in space was a question at the forefront of many, if not most, space enthusiasts' minds: would they personally get the chance to ride to space aboard the shuttle? NASA had strived to make people feel like part of the Mercury, Gemini, and Apollo programs through its open information approach, but it had not made promises to provide access to space to all. As the 1970s rolled around, achieving personal goals and relying on one's own experiences as a means of understanding became important to increasing numbers of Americans. One woman, Mary Manning, expressed in a letter to the editor of the *Los Angeles Times* that "we, the public" got tired of NASA's "self-serving media campaigns." Manning asserted: "Give us something that offers a genuine sense of national effort and participation in a great enterprise, a real flow of information and involvement, and we all might again get interested."[48]

As part of its effort to cultivate space supporters, NASA strived to show it was working to open space flight as widely as possible. NASA and Nixon had touted the imaginary of the shuttle as a vehicle that would expand participation in space flight. Agency officials consequently promoted the shuttle's ability to democratize the ranks of space flyers and accommodate people for a variety of purposes and with relative ease, in contrast with the rigors of space flight associated with the Mercury, Gemini, and Apollo capsules. As depicted in a NASA publication released in 1971, when the agency still linked the shuttle with a space station: "For passengers, a shuttle trip to or from space stations may be similar to a business trip by air to a distant city. . . . While the astronauts pilot the craft, the passengers will relax in comfort comparable to flight in today's airlines." Shuttle flights would become so routine that these vehicles might "fly into space on timetables like those of buses, trains, ships and airliners."[49]

NASA highlighted the fact that the shuttle's reduced forces of acceleration during launch and reentry would mean that many people would qualify for space flight. "Space flight will no longer be limited to intensively trained, physically-perfect astronauts," stated a 1977 Johnson Space Center brochure, which noted that the more benign flight conditions "will welcome the nonastronaut space worker of the future." Astronaut Deke Slayton told a Florida newspaper that the shuttle's flight similarities to a commercial jetliner meant that it could be possible to fly anyone. Jim Bilodeau, director of shuttle crew training at Johnson Space Center, told the *New York Times*, "Basically, we'll be able to take everybody but the walking wounded." A NASA fact sheet from 1985 claimed that the acceleration forces were comparable to those experienced "on some carnival rides." Moreover, the agency indicated that shuttle flyers would come from a range of backgrounds. One 1972 publicity document noted: "They may be scientists, engineers, technicians, journalists, television crews or others whose business takes them into space. As experience increases the assurance of safety, men and women of many organizations and many countries will be among the passengers." Beggs asserted, "If this program is going to be what we hope it will be, we want a very broad spectrum of people involved in it. We don't want just scientists and engineers."[50]

Indeed, imaginative types at NASA were thinking about the prospect of the shuttle opening space flight to ordinary citizens. In March 1970, NASA public affairs officers sent to magazine editors a sample article of a new NASA news subscription service that noted that "the fact remains that you and I, with no more qualifications than now required to fly on a passenger jet, will one day fly on a space shuttle—for a price of course." It quoted NASA deputy associate administrator for manned space flight Charles Matthews as saying: "We can expect average people to fly into space, to visit, to live and work," with the shuttle operating akin to a commercial airline service for passengers and cargo. The article estimated that NASA would be able to offer fifty round trips per year costing passengers $5,000 each. The shuttle would "operate as a common carrier," according to a NASA brochure, "serving essentially anyone who can buy a ticket or pay the freight cost."[51]

It is perhaps unsurprising, then, that many journalists and citizens began to believe, as a *National Geographic* article characterized it, that the shuttle seemed to mark "the beginning of the people's space program." NASA's effort to promote an exciting future for the shuttle inspired the public in ways for which the agency was not fully prepared. As the shuttle's operational phase approached, NASA began to receive letters from people all over the country and the world asking to fly aboard the shuttle, many having been led by NASA and media rhetoric to believe that the agency was taking reservations for passengers.[52] As I will reveal in subsequent chapters, NASA officials, guided by the new imaginary for human space flight, would respond to the enthusiasm by establishing means for members of the public to participate directly in shuttle missions.

Linking the Imaginary with Traditional American Themes

Human space flight began in the United States as a quest to demonstrate the nation's technological and ideological superiority over the Soviet Union. The shuttle shifted the narrative of human space flight from a geopolitical race between superpowers to the pursuit of benefits and accessibility to space for many. Even so, NASA's original identification of human space flight as a symbol of national pride and strength ran deep for many Americans. Therefore, just as they had done in the Mercury, Gemini, and Apollo eras, NASA officials associated the shuttle with quintessentially American, forward-looking themes including new beginnings, the pioneering of new frontiers, social and technical progress, and a collective sense of American ingenuity and propensity to dream. This time, however, the agency would connect the themes with the opportunities that the new vehicle would extend to Americans.

Founded by those seeking religious freedom, resources, and other forms of opportunity, the United States has historically self-identified with new starts. NASA and others invoked this ideal with great frequency in discourse about the shuttle during its development and early operational phase, indicating that the new vehicle represented a transition to a "new era" in space flight. Americans have long been fascinated by their transportation systems, and NASA applied the phrase "new era" to the shuttle as the next stage in the evolution of the nation's technological conveyances. NASA claimed in 1975 that a "whole new era of transportation will come into being" with the shuttle's advent. A 1977 Johnson Space Center publication likewise asserted: "Building upon previous achievements, new plateaus in air and space transportation have been reached—military aviation, airmail, commercial passenger service, the jet age, and manned space flight. Now a new era nears. The beginning of regularly scheduled runs of NASA's Space Shuttle to and from Earth orbit in the 1980's [sic] marks the coming of age in space."[53] Merging the ideas that the shuttle would be routine and exceptional all at once, Grumman referred to the shuttle in 1971 as a multipurpose "Space Truck" capable of carrying a variety of payloads to orbit. Kennedy Space Center offered visitors a similar juxtaposition of the ordinary and the extraordinary in stating that the shuttle would "reenter the atmosphere regularly, landing like a sensational jetliner."[54] The agency propagated the image with colorful brochures, such as one illustrated by artist Robert McCall, the cover of which featured an animated shuttle zooming through space with painted stripes and a large number 3 on the tail and wing, reminiscent of a race car (see fig. 3.2).[55]

NASA and its counterparts promoted the idea that the shuttle heralded a new era in still other ways that reinforced the imaginary of the vehicle as a democratic technology. A 1977 Rockwell publication focused not on the novelty of the shuttle per se but on its benefits for Earth and its inhabitants. *Space Shuttle: A Promising New Era for Earth* enumerated ways in which space exploration could be used to help solve societal problems. It stated: "We long have had the technology with which to do this work; now we have the transportation system: Space Shuttle, a versatile vehicle that will permit us to carry out these and numerous other useful activities." Fletcher, meanwhile, used the phrase when the orbiter *Enterprise* rolled out at Rockwell's Palmdale, California, facility in 1976. "This day, we're about to enter a new era," Fletcher said, noting that the shuttle would open space to "all people" and be "the beginning of a system of space transportation in which we will enter the environment of space permanently."[56] Kennedy Space Center referred to STS-1 (the name an

FIGURE 3.2. The cover of a 1972 NASA brochure depicts an artist's conception of a space shuttle racing through space, connoting that the vehicle's travels would be routine and exceptional at the same time. Brochure from NASA.

abbreviation for Space Transportation System), the first shuttle mission aboard the orbiter *Columbia*, as marking "the dawning of a new era in the exploration and utilization of space" and explained that "unlike spacecraft of the past, the Space Shuttle qualifies as a true spaceship, a reusable vehicle affording routine access to space."[57] Upon the successful return of STS-1, President Ronald Reagan declared "a new era in space travel."[58] In each of these usages, NASA and those promoting the shuttle implied that the new vehicle was distinct from earlier human space flight endeavors and of great importance for the nation. They simultaneously framed the new vehicle in down-to-Earth, pragmatic terms and conveyed its excitement and promising tomorrows.

A related theme the agency invoked from the pages of Americana was the idea that the new vehicle would help open the frontier of space, just as hardy souls had pioneered the American West. In doing so, NASA and the shuttle would contribute to the nation's social and technological progress. The metaphor was not new to NASA: President John F. Kennedy and NASA officials had relied upon it in discussing the nation's move into space in the early 1960s. As a 1981 *New Yorker* piece said, "People at NASA love to draw analogies between the development of space and that of the American West; they say that the Shuttle is like the covered wagon, which started out hauling a few simple goods and wound up settling a whole country." Alan Ladwig, who held a variety of positions within NASA to engage the public with the shuttle, recalled colleagues around the agency referring to the shuttle as the "prairie schooner of the future."[59] On the eve of STS-1's launch, von Puttkamer compared the shuttle to the railroad's role in opening the western frontier of the United States in that the new vehicle would enable economical and easy access to space to build large structures and eventual colonies.[60]

NASA also publicly identified the shuttle with progress and forward movement by emblazoning a new logo on the left wing of each orbiter and on the right breast of the astronauts' flight suits. It also adorned all NASA letterhead, publications, and other communications materials. The effort began with a directive from President Nixon and carried forward by President Ford for the National Endowment for the Arts to review graphics used across federal government agencies and improve their quality for

FIGURE 3.3. NASA's original "meatball" insignia (left) was replaced in the 1970s by the "worm" logo, which was intended to modernize the agency's image and connote a sense of movement into the future. Insignia from NASA.

more effective communication with citizens. The National Endowment for the Arts concluded that NASA's original blue "meatball" insignia did not reflect "the most highly technological, exciting, and contemporary agency in the Federal Government." NASA thus scrapped the insignia that had been associated with its previous human space flight feats and adopted a new logotype: a stylized rendering of the letters *N-A-S-A* in bright red (see fig. 3.3). Fletcher believed that the new, warm color of the logo would counter the "cold, mechanized, technological image sometimes associated with NASA" while offering "a feeling of unity, technological precision, thrust and orientation toward the future."[61]

In addition, the idea that an indefatigable American spirit gave rise to the shuttle appeared frequently in officials' public discourse, particularly when comparing American capabilities to those elsewhere in the world. Beggs remarked in welcoming *Enterprise* home after the test article was taken on a multicountry tour that "the drive, the ingenuity, the determination, the imagination that has always characterized the American pioneer spirit is built into this craft and into all the Shuttle orbiters." On some occasions, allusions to continued competition with the Soviet Union crept in, as officials harkened back to a sense that America's unbridled perseverance stemmed from the nation's commitment to democracy. Upon the landing of STS-1/*Columbia*, Reagan declared, "Today our friends and adversaries are reminded that we are a free people capable of great deeds. We are a free people in search of progress for mankind, and today we found a little more." General James Abrahamson, associate administrator for space flight in the early 1980s, asserted that the shuttle was "a fantastic comment on the American spirit" and symbolized that "drive which wells up in free people, in Americans, to win, to strive for victory over others or against our own goals, to be first, to be number one."[62]

Finally, NASA along with President Reagan sought to accentuate the notion of the shuttle as a democratic technology by linking the vehicle with Americans' hopes and dreams and suggesting that those aspirations were shared by the nation's citizens. Abrahamson expressed at a meeting of journalists that the shuttle was a "dream" for students, businessmen, scientists, and journalists alike—a dream that "belongs to every one of us." He said that "what the Space Shuttle really is all about is a means of making the dreams of Americans come true." STS-1 pilot John Young remarked, following his historic mission's landing, "The dream is alive." Reagan told Congress that the start of the shuttle program "did more than prove our technological abilities. It raised our expectations once more. It started us dreaming again."[63]

The agency further strived to conjure a sense of unity and collective ownership of the shuttle. A promotional brochure for Kennedy Space Center from the late 1970s was titled *Welcome to Your Spaceport*. Abrahamson told *USA Today* in 1984, "The way I feel about it—and I think most people in the agency do—is that the Shuttle program really belongs to the American people. This is an important step in making them really understand and feel what that ownership is." NASA's first Black astronaut to fly in space, Guion Bluford, also suggested a common possession of the shuttle. Upon returning from his first mission, STS-8, in 1983, he said: "I feel very proud to be a member of this crew, and I think we have a tremendous future with the space shuttle, I mean all of us."[64]

NASA's words resonated with many Americans. Some people were so moved by the first flight of *Columbia* that they sent letters and poems to NASA praising the mission's success. Others, feeling empowered to help the shuttle's cause, wrote in with suggestions about how NASA could make the program more productive. People's pride in the shuttle showed in their willingness to let NASA know their concerns down to the aesthetic details of the vehicle. When NASA decided to stop painting the external tank white after the first two missions to save on cost and weight, leaving it in its natural rust-hued state, many wrote to express their displeasure. Shuttle official James Odom remarked, "You wouldn't believe the ugly letters I [received] when we took the paint off. 'That old ugly colored tank.' Most of them were from ladies that just thought it really looked good before." Odom added: "I wouldn't have thought the public would have paid much attention to it."[65] These reactions helped to substantiate NASA's continued use of discourse characterizing the shuttle as a vehicle owned by the nation's citizens.

Proving Capabilities, Competence, and Credibility

When NASA and the Nixon administration rolled out plans to develop the shuttle as the nation's next major human space flight initiative, they registered a very tall order, technically speaking. While the agency and its aerospace industry partners had built systems capable of sending humans into deep space, the shuttle represented an entirely new approach to space flight. The orbiter would need to accommodate and manipulate heavy, bulky payloads. It would need to sustain large crews. It required a new propulsion system for launch and on-orbit maneuvering as well as a thermal protection system to withstand atmospheric heating upon reentry. It would need to be outfitted for ground landings rather than the ocean splashdowns of the past. And to meet promises that the shuttle would enable routine, less costly access to orbit, NASA would need to service the orbiter rapidly for reuse.

Even if people doubted human space flight's value, NASA's success with the Apollo program had earned it a reputation as a can-do organization, capable of tremendous technical achievements. The agency's experience with Skylab had demonstrated that humans could live and work for long periods in space. At the same time, Skylab's experiments to produce perfect ball bearings and crystals in microgravity did not yield the results NASA had hoped for to revolutionize manufacturing processes and the computer industry. Thus, the stakes were high for the agency to prove that the shuttle would make space flight routine, economical, and critical to advancing commercial and industrial applications in orbit. Preserving an image of credibility and competence was also essential for NASA to gain support to build the space station that was still part of its long-term aspirations.

NASA's continued messaging about the shuttle's ability to deliver societal benefits and to create an exciting future in space throughout the 1970s intrigued many Americans. But as NASA moved into developing the shuttle system and producing the orbiters, technical challenges plagued the agency and threatened the 1979 target launch date for the first fully functional orbiter, *Columbia*. Some of the reusable main engines exploded during testing, and NASA drew down the shuttle program's funding reserves trying to address this and other problems. Congress granted the agency additional resources to preserve the launch date, but additional setbacks would preclude the shuttle from being ready to launch that year. Of most serious concern, many of the orbiter's protective ceramic heat tiles had fallen off when Rockwell transported it from its assembly plant in California to Kennedy Space Center atop a Boeing 747, and technicians struggled to reaffix them properly. They ultimately got the job done, but the issue revealed that operating the new vehicle was going to be more complicated than the agency had suggested.

Part of the reason for the technical challenges and cost overruns was that NASA had trimmed the shuttle's development program, introducing more risk when it accepted Nixon's condition that the shuttle be developed for a fraction of the agency's initial projected cost. Some citizens were sympathetic and even tried to help: NASA received many well-intentioned letters from people across the country who offered ideas to solve the shuttle's thermal protection system difficulties as well as to maximize the cost-effectiveness, efficiency, and safety of shuttle operations.[66] Officials, however, were unsettled by a preponderance of critical comments directed at the agency. Some members of the space science community denounced the shuttle's cost growth when their own programs became targets for solving the shuttle's budget woes. The General Accounting Office blasted the shuttle program in 1977, questioning its technical integrity, environmental impact, and ability to yield the demand and lower launch costs that NASA had claimed.[67]

Many members of the media were particularly vocal in airing the issues besetting the shuttle. Journalists derided it, referring to it as "The Plane That Won't Fly," "America's Space Shuttle Lemon," "Aluminum Dumbo," and "The Spruce Goose of Outer Space." Bob Thompson, manager of the shuttle program at Johnson Space Center during the 1970s, recalled having to defend the shuttle on a 1978 episode of *Nightline* against a very antagonistic Jules Bergman. Thompson thought Bergman was unfair in calling out the vehicle's cost growth using unrefined cost estimates from 1972.[68] Others blamed the shuttle delays for causing the abandoned Skylab to fall, uncontrolled, out of an unstable orbit to land wherever it might on Earth in 1979. Originally the agency had planned to have the shuttle ready in time to boost the eighty-ton laboratory into a stable orbit or deorbit it into the Pacific Ocean in a controlled manner.[69]

NASA officials wanted to turn around the opinions of the news media to show American citizens that the shuttle was reliable, capable, and valuable. In the 1960s the agency had offered members of the news media open access to its activities so that they would be knowledgeable about the space program and report positively about it. That effort had fallen by the wayside during the mid-1970s with public affairs personnel changes. Duff, who was appointed as NASA's new public affairs chief when negative stories about the shuttle were emerging, wanted to restore its communications machinery to "how it was during the Apollo days." According to Duff, "the attitude was basically that the best defense was a good offense, that you should always stay ahead of the problem instead of behind it. You should anticipate things and meet them, rather than having them catch you and having to react to them." Duff's solution was to enhance NASA's transparency and outreach to the media. He told his staff, "You're going to see more newsmen in the next six weeks than you've seen in the last two years . . . you've got a story to tell, and we're going to make you so available the media is going to be tired of talking to you."[70]

Duff recommended to administrator Robert Frosch in 1980 that the agency support an "STS Continuing Information Program" to "convey more adequately" the shuttle's breadth, complexity, and long-term significance. Duff wanted to make sure NASA remained in control as the authoritative source for information about the shuttle and to expand media coverage to a broad spectrum of subjects, not just development challenges. The idea was, Duff said, to portray in "understandable incremental segments to a variety of audiences" the shuttle's story, which was "not the story of a single flight or a machine" but that of "a 'leapfrog' technology which promises a new Space Era." To do this, he said, NASA needed to tell reporters and others candidly about the issues it faced along with its progress in developing the shuttle. He recognized the primary audience as the news media but envisioned public relations materials about the shuttle being useful in dealing with any external constituency, including officials in the White House and Congress as well as potential shuttle users.[71]

The public affairs office under Duff started weekly press briefings, bringing in different experts to cover various aspects of the program. They printed transcripts of the briefings and made each one a chapter in what the agency dubbed the Shuttle Press Manual. By conveying to press members how complex the shuttle program was and how NASA was working to address the technical challenges, the public affairs office was, Duff explained, "trying to get the media to get over the idea that we were afraid of them, and that we were afraid of the program."[72] Many of the journalists with whom they engaged were new to covering the space program, and Duff and his team hoped that their efforts would lead to more stories that reflected positively, or at least empathetically, on the shuttle.

The approach seemed to be effective, as the agency established a rapport with reporters and created a press corps for the shuttle akin to the Apollo press corps of the 1960s. ABC, CBS, NBC, and CNN all planned live coverage of launch, reentry, and landing of STS-1.[73] Lisa Malone, a public affairs official at Kennedy Space Center, recalled that the shuttle became the NASA program of highest interest to reporters, who had "a thirst to know all of the technical details about it for many years." In January 1982 Duff shared with deputy administrator Hans Mark his observation that the media tide had turned in NASA's favor a few months before the STS-1 mission. The shuttle dominated a number of magazine covers in 1981 and was cast in the role of a "national triumph."[74] Following STS-1's landing, newspapers lauded the shuttle's capabilities and praised NASA for remaining a "can-do" agency.[75] With coverage positive on the whole, Duff informed Mark that NASA "enjoys a remarkable reputation for competency and credibility."[76]

NASA's public affairs office continued to nurture its relationship with the news media as shuttle missions got under way and the agency sought continued visibility and support for the vehicle. The public affairs office compiled a handbook advising journalists about how to cover shuttle missions and followed with a general shuttle news media handbook.[77] NASA made viewing accommodations for domestic and foreign press agents at launches and landings as well as at mission control at the Johnson Space Center. News media received access to live audio and video feeds of the launches and missions as well as stock film footage. Beginning in 1983, NASA initiated a service in which members of the media could dial a phone number to listen to communications between mission control and the shuttle crews in orbit and also began to conduct space-to-Earth press conferences. The agency occasionally permitted journalists to get even closer. *New Yorker* writer Henry S. F. Cooper lived and worked with the STS-41G crew before their October 1984 flight to capture the "human dimensions" of training.[78] The only exception to NASA's openness was when a shuttle mission flew a defense payload and the agency withheld information about the mission and payload per an agreement with the Department of Defense.

NASA officials and their industry partners took care to highlight their performance victories as flights continued to garner media, public, and political validation of the program. For one, the agency declared the shuttle "operational" in June 1982 after the first four missions, even though the agency had initially planned for six test flights. The next month, Rockwell released a brochure stating, "Space Shuttle is becoming more glamorous, and those who are building and operating it couldn't be more pleased. So successful are its orbital flights getting that they generate little suspense. The Shuttle orbiter is taking the risk out of space operations, as it was designed to do."[79] Astronauts endorsed the shuttle's capabilities, with STS-3 commander Jack Lousma declaring that *"Columbia performed magnificently"* and pilot Gordon Fullerton calling the vehicle "an unbelievably beautiful flying machine."[80] Following the maiden flight of *Challenger* in 1983, General Abrahamson told the *Washington Post*: "It was a great mission, incredibly routine, which is what we want."[81] Abrahamson pointed out in congressional testimony and speeches that payload owners felt confident in NASA's performance, and shared the numbers of shuttle customers NASA enrolled, including those they "stole" from France's Ariane, their major competitor for satellite payloads. Abrahamson also reported reductions in the number of anomalies encountered with each passing shuttle mission.

NASA aimed to demonstrate accountability for delivering on its promises concerning the shuttle in still other ways. Through news releases and mission status briefings, NASA kept press members and in turn the public informed about plans for scheduled missions as well as the details of missions as they unfolded. Such details included satellite deployments and retrievals and the achievement of other key mission objectives. Further, the agency strived to instill public confidence in the shuttle and NASA more generally by continuing to maintain an open and responsive program on many fronts. NASA's public affairs offices busily fielded public inquiries, which surged around the time of shuttle launches. Kennedy Space Center's Malone attested to the huge amount of "fan mail" and other forms of contact NASA received and responded to during the early shuttle era. She recalled: "We had so many phone calls I would go home and my ear was raw. . . . Some public relations folks are always trying to throw the fishing line out and reel them in. [In NASA's case] the fish were jumping on us."[82] NASA would reply to citizens' queries about the shuttle with fact sheets, photographs of crews, and mission patch stickers. The agency addressed the curious, skeptical, and uninformed alike by sharing with them publications about daily life and activities aboard the shuttle, ranging from mission tasks to what the astronauts ate and how they went to the bathroom in space.[83]

Even after declaring the shuttle operational, NASA recognized that it still had a way to go in fulfilling the agency's vision of a routinely and broadly used vehicle. In 1983 Beggs acknowledged at a House appropriations hearing that the shuttle was pushing the state-of-the-art as hard as, if not harder than, the Apollo Saturn V had.[84] NASA leaders were aware, too, that the vehicle remained experimental in many ways and was not yet on a trajectory to becoming cost-effective. Accordingly, NASA kept up its efforts to publicize its plans and successes, always working to convey that the shuttle's progress in fulfilling the imaginary of an accessible and purposeful vehicle overshadowed the technical hitches and delays. Part of that strategy included going beyond messaging and bringing the nation's people as close to the new vehicle as possible.

4

Inviting Virtual Participation

Being able to articulate the shuttle's value and show that it had developed into a highly capable vehicle was a key piece of NASA's ability to legitimize the new space transportation system to the American people. But the completed shuttle also was critical to NASA's public engagement efforts in other respects. Public attention during the Mercury, Gemini, and Apollo eras had come from NASA having a visibly active flight program that offered visual and sensory experiences. The agency had provided citizens opportunities to learn about space flight missions directly from the astronauts, to visit its facilities to see firsthand the technology that made those journeys possible, and to take in the spectacle of a launch. These opportunities let people develop a different sort of appreciation for the space program than reading or hearing about NASA's progress and plans permitted. "Once exposed to the magnitude, excitement and sense of an adventure that comes with close-up observation of space exploration activities," one NASA planning document for accommodating visitors at shuttle launches read, "the individual, no matter what his station in life, has become involved." The document described an "emotional experience" and an "intellectual broadening" that came after one had "seen the hardware, visited the sites, met the people, sensed the magnitude of it all, and participated in the action."[1]

In conjunction with its messages about the shuttle's value, NASA aimed to realize the vision of a democratic space vehicle and make all Americans feel like participants in the shuttle program by allowing them to engage with the new vehicle and its missions in a range of ways. Agency officials hoped that "experiencing" the shuttle would generate enthusiasm among members of the public about the significance of the agency's new initiative, in turn making them supporters of human space flight. In this chapter I show how NASA aimed to make Americans feel connected to the shuttle program once the vehicle began flying. The agency offered tours and overflights of the shuttle test article *Enterprise*, welcomed citizens to shuttle launches and landings, harnessed communications technologies, and provided opportunities for people to experience the sights, sounds, and daily life aboard the shuttle while keeping their feet on the ground. All drew on the shuttle's unique features—its relative roominess and ability to accommodate larger crew sizes and longer mission lengths—as well as the availability of new communications technologies. In doing so, NASA reinforced the notion of a spacecraft intended to serve all Americans and gave millions the chance to view and experience the shuttle in more proximate ways than the agency had made possible during the Apollo missions.

It is worth noting that the agency's employment of these approaches did not follow a master plan. While considering some audiences early on, NASA officials, from the administrator to public affairs specialists to those in other offices, often made decisions about whom to engage and how as ideas and opportunities arose and as challenges and critics became evident. Different organizations and individuals within NASA often held different opinions on what approaches and what audiences should be pursued, and officials had to convince one another of the merits and risks of various public engagement choices. Through these public engagement efforts, NASA found allies who valued the shuttle, many of whom in turn became integral to NASA's efforts to further promote and sustain the program. The American public and the shuttle thus co-evolved, with these public engagement efforts helping to shape interest in the shuttle program as much as these offerings and public participation in them affirmed the shuttle as a democratic technology belonging to the nation.

Introducing *Enterprise*

Historian David Nye has noted that Americans have an affinity for "the technological sublime," associating machines they have developed to shape their environments and destinies with a sense of awe, power, and national greatness. It is this trait, Nye argues, that has drawn citizens to witness and experience events ranging from world's fairs to rocket launches. But more than just wanting to see a launch of any space vehicle, people tend to be drawn to seeing spaceships with the capacity to carry people. Public and media attendance at launches of NASA's scientific probes have always paled in comparison to the spectatorship garnered by human missions. NASA administrator James Beggs attested, "The plain facts are that when we are flying men and women in space, it has a huge impetus to the interest that the public has in the program." Kennedy Space Center's Lisa Malone observed that human space flight has "a love potion—I don't know what you call it, but it's got a lure that is attractive to the public. It's magnetic in a way that you just can't describe."[2]

Millions had flocked to peer at the Mercury, Gemini, and Apollo capsules as the agency took them on national and international viewing tours. Likewise, even in the face of the zeitgeist of pragmatism, NASA found that many Americans were eager to at least catch a glimpse of its new technological marvel, the space shuttle. This time the awe was somewhat differently placed. In the 1960s, people could look at NASA's space capsules and behold the ingenuity that had enabled Americans to reach a new world. Now they could lay eyes on a space plane: a craft that, with the physique of an airplane, made the seemingly fantastical, abstract, and unobtainable appear to be suddenly familiar and within reach. It captured the imagination and suggested the real sense of service that NASA was promising all at once. James Hartsfield, a Johnson Space Center public affairs officer, noted that the shuttle's technical attributes were practically "tailor-made" to intrigue citizens. "People love wings," he said. "Engagement with the shuttle was easy in that respect. People liked it

FIGURE 4.1. NASA officials and members of the *Star Trek* television series cast celebrate the rollout of the shuttle test article *Enterprise* in Palmdale, California, on September 17, 1976. Standing from left to right: NASA administrator James Fletcher; DeForest Kelley (Dr. "Bones" McCoy on the series); George Takei (Mr. Sulu); James Doohan (chief engineer Montgomery "Scotty" Scott); Nichelle Nichols (Lt. Uhura); Leonard Nimoy (Mr. Spock); series creator Gene Roddenberry; US Rep. Don Fuqua; and Walter Koenig (ensign Pavel Chekov). Photograph from NASA.

from the start. People wanted to see something fly home, and they wanted to dream about a way they could get on an airplane and fly to space."[3]

September 17, 1976, marked the start of NASA's effort to afford the public visual and virtual access to the shuttle. On that day, NASA debuted the first shuttle orbiter, *Enterprise*, at Rockwell's assembly facility in Palmdale, California. The rollout ceremony itself was open to invited government leaders, aerospace and other business executives, media representatives, and other VIPs, including cast members of the original *Star Trek* television series after whose spaceship the new orbiter had been named (see fig. 4.1). An air force band played the *Star Trek* theme song as *Enterprise* emerged from her hangar. Finally able to point to the embodiment of a spaceship that had for several years been only a concept, Rockwell chairman Willard F. Rockwell Jr. told the crowd that "those of little vision who accuse the space program and particularly the shuttle of being too far ahead of its time" could now be refuted.[4] The next day, some thirty-five to forty thousand people showed up to an open house that Rockwell the company had planned for its employees. The media had announced it as a public event, and Rockwell quickly arranged to accommodate the enthusiastic and inquisitive hordes of visitors. When NASA and Rockwell transported *Enterprise* from Palmdale to the agency's Dryden (now Armstrong) Flight Research Center at Edwards Air Force Base in California's Mojave Desert in January 1977, citizens lined the roads along the thirty-five-mile stretch to witness the spectacle.

NASA provided numerous opportunities for the public to see the new orbiter during the course of its testing program. Throughout 1977 the agency conducted approach and landing tests (ALT) at Dryden to evaluate *Enterprise*'s aerodynamic and flight-control characteristics. *Enterprise* was mounted atop the modified Boeing 747 carrier aircraft that would be used to transport the orbiter across the country from landing to launch site. The ALT protocol included tests to appraise the flightworthiness of the combination, first without a crew or using power and then with astronauts powering the orbiter's systems. Starting in the summer *Enterprise* flew five free flights, touching down on the dry lakebed landing strips at Edwards, which was one of the shuttle's designated landing sites. The agency planned carefully for invited special guests and members of the public who requested vehicle passes to view these tests onsite. In March 1978 *Enterprise*, with its carrier aircraft, flew to NASA's Marshall Space Flight Center in Huntsville, Alabama, for vibration tests. When it stopped in the Houston area for refueling, some 240,000 people came out to catch a glimpse of the shuttle after NASA's Johnson Space Center issued a press release announcing the shuttle's stopover at Ellington Field. Thousands more were on hand to greet *Enterprise* upon landing in Huntsville. A

NASA official on hand to show people *Enterprise* during its stop at Ellington Field said, "This is different than Apollo and going to the Moon. I
think the Shuttle is coming closer to the people. It is something they can
relate to. They wanted to know when they can go on it."[5]

NASA had built *Enterprise* as a test article for the shuttle program, lack-
ing engines or thermal protection, but the agency had intended to refit
it for space flight. When the orbiter's fuselage and wing designs changed
during the construction of *Columbia*, however, the agency scrapped plans
to make *Enterprise* functional, as the required overhaul would prove very
costly. While *Enterprise* would remain an important reference asset for
the shuttle program, NASA recognized an opportunity to showcase the
vehicle domestically and internationally. As early as 1978, NASA officials
contemplated the prospect of taking *Enterprise* to the Paris Air Show, the
world's foremost aviation and aerospace event. Although the feat was
physically possible, NASA retrenched from pursuing it when shuttle
program director Myron Malkin balked at the potential for damage. Why
risk an accident or a terrorist attack on the shuttle carrier aircraft or *Enterprise*, he asked, for a "clearly non-essential public relations gesture?"[6]
Beggs rebuffed the idea as well when it resurfaced through public affairs
chief Brian Duff in the early 1980s. The Central Intelligence Agency, the
Department of Defense, and the State Department also objected. Some
members of Congress, meanwhile, favored the idea of promoting American technology in an increasingly competitive international aerospace
and high-technology market. Plus, the pilots of the 747 believed that the
flight would be easy enough. According to Duff, "Once Beggs' mind began
to grasp what would happen if we did this successfully, he was all for it"
and overrode the objections of the other agencies.[7] NASA made a plan to
go to the 1983 Paris Air Show.

There was one problem: the agency had a difficult time convincing
French aviation authorities to allow demonstration flights. The French
attributed their hesitation to fatal accidents at the air show in recent
years, but the idea of Americans touting their technology over Parisian
skies when the French were heavily marketing their Ariane launch vehicle likely added to the resistance. The mayor of Paris, however, ultimately gave approval for NASA to fly *Enterprise* around the city's Boulevard

Périphérique. Parisians were in awe of the spectacle. NASA associate administrator for space flight James Abrahamson recalled enthusiastic spectators calling out "Navette! Navette!" (Shuttle! Shuttle!) as *Enterprise* flew overhead. NASA public affairs officer Louis Parker reflected: "The traffic in Paris is bad anyway, but every time the orbiter was flown around—and it would fly [at] 1,000 feet, 1,500 feet, very visible—it just would create all sorts of havoc. As a matter of fact, that particular year, the French Open was happening at the same time as the Paris Air Show. John McEnroe was playing. He literally stopped a tennis match one day because it was flying across."[8]

The plan to attend the Paris Air Show ended up turning into a month-long multinational tour. When other European nations found out that *Enterprise* would be nearby, they asked if the shuttle could come to them. Parker recalled, "Everybody wanted to have it come to their airport. And . . . if it couldn't come to their airport, could they at least fly over our airspace and drop down so people could see it?" So the agency built a tour route that also took *Enterprise* to four other Western European countries and Canada. Everywhere, it was met with hundreds of thousands of enthusiastic viewers. In addition, en route there and upon its return, it stopped in several American cities, including Colorado Springs, Wichita, Dayton, New York City, and Washington, DC. In the nation's capital it flew low and majestically around the I-495 Beltway and up the Potomac River. As it flew, *Enterprise* commanded oohs, aahs, and applause from the hundreds of thousands of onlookers who witnessed the spectacle. Tens of thousands more were present when the shuttle landed at Dulles International Airport, and NASA outfitted them with small American flags and shuttle lapel pins to accentuate the occasion's importance to the nation. "This stopover," Beggs told the *Washington Times*, "will allow the taxpayers to see what some of their money has gone for. We're very proud of the Enterprise and the whole shuttle flight system. We want the American public to be proud, too." Indeed, NASA curried the sense of appreciation it had hoped the tour would elicit. "That's the most beautiful thing I've ever seen," one local woman told the *Washington Post*. A viewer who had come from Atlanta to see the shuttle exclaimed, "It's worth every penny even if it does nothing but raise the morale of the country."[9]

NASA made plans to showcase *Enterprise* at the world's fair in New Orleans the following year. Beggs noted, "We at NASA are very much aware of the potential benefits for the agency, for the Space Program and for our nation which can result from exposure to an audience of the size and enthusiasm which the fair will provide."[10] The agency flew the shuttle aboard the 747 to Mobile, Alabama, where the orbiter, too large to reach the fair site by rail or road, was loaded onto a barge for transport to New Orleans.[11] A multiday stopover in Mobile brought thousands of school-children to see *Enterprise*, and the agency accommodated them with close-up encounters. Parker recalled: "The kids would literally walk by it. There was even a school for the handicapped that came; I think the kids were blind. They wanted them to literally be able to touch it. There was a picture of me that appeared in the Mobile newspaper . . . holding up a little blind kid to touch the nose of the Enterprise. That was really kind of neat." During *Enterprise*'s six-month stay in New Orleans, some 2.5 million fair visitors viewed the orbiter. They also saw a full-scale mockup of the shuttle orbiter flight deck and Spacelab laboratory module as well as a video production about shuttle launches and missions.[12] The experience of beholding *Enterprise* showed people the reality of the new human space flight adventure being set into motion.

Extending Opportunities to View Launches and Landings

As the shuttle program became the mainstay of human space flight in the 1980s, NASA continued to welcome hundreds of thousands of people annually to its visitor centers. Seen by the agency as a means of both "satisfying the public's interests in our programs" and "showing the taxpayer what we do with their tax dollars," visitor centers fulfilled the agency's aims of remaining accountable to citizens while improving their understanding of its facilities and programs to build their support. The visitor centers at Kennedy and Johnson featured exhibits on the shuttle. Johnson changed its exhibit regularly to reflect cargo and experiments being flown on the latest shuttle mission.[13] Kennedy added the shuttle launchpads to its standard visitor tour, while Johnson allowed guests to view the shuttle simulator astronauts used for training purposes. Attendance at both centers burgeoned with the shuttle's arrival, bringing one million visitors annually to Johnson by the 1980s and pushing visitation to Kennedy to more than two million. Both visitor centers underwent major expansions during the 1980s and 1990s to accommodate the new wave of interest.

Once the shuttle began flying missions, NASA was able to engage Americans through the power of witnessing in person the launch of the new space transportation system. The agency did not really need to persuade citizens to make the trek to see a launch. Millions of people who had witnessed the Mercury, Gemini, and Apollo takeoffs were already aware that attending a launch was not just an act of observation; it was a dramatic multisensory experience whose magnificence was like no other. As an ad in the National Space Institute (NSI)'s *Space World* magazine promoting the organization's shuttle launch tours noted: "You've seen it on TV, but nothing can compare to being right there—inside the Kennedy Space Center gates—when six million pounds of thrust propel America's reusable spacecraft into orbit . . . and you feel the vibration . . . hear the roar . . . see the brilliance!"[14]

NASA had begun receiving requests from citizens to view STS-1 from onsite at Kennedy Space Center since *Enterprise*'s approach and landing tests. As it had during the 1960s, NASA orchestrated a massive effort to accommodate the requests and visitors. The "guest operations" arm of the agency's public affairs office handled arrangements for officials from the White House, Congress, other federal agencies, and foreign governments as well as the invitees of high-ranking NASA leadership, the mission's astronauts, and shuttle contractors and customers. NASA issued thousands of "car passes" to members of the public that allowed holders to drive onsite and watch the launch from designated locations. Tour buses leaving from Kennedy Space Center's visitor center transported thousands more to viewing sites around the premises. The agency's public affairs officers broadcast live commentary over loudspeakers and radio to explain the developments leading up to the launch and in the moments following it, putting the engineering jargon of mission control into accessible terms. The agency made special shuttle souvenirs available for attendees to buy. Popular with stamp collectors and others looking for a unique memento to commemorate the day were envelopes bearing an

FIGURE 4.2. NASA issued this postal cover honoring the June 18, 1983, launch of STS-7/*Challenger*, the first US space mission to include a female crewmember, Sally Ride. Image from NASA.

image of the mission's patch or a related picture and some information about the mission inside. Purchasers could stamp and address the envelopes, which would be cancelled on the day of launch and mailed to the designated recipients (see fig. 4.2).

Some 80,000 people from all walks of life assembled at Kennedy Space Center—the largest crowd ever gathered there to date—to view the launch of STS-1 on April 10, 1981. Among them were nearly 3,500 reporters, photographers, and other members of the media. Because the launch was scrubbed due to a computer malfunction and did not happen until the following day, the total crowd onsite to see it dropped to 45,000.[15] Another one million or so saw the launch from the surrounding area. The agency delighted in the reactions of those who witnessed the historical event, which seemed to erase all traces of doubt about NASA's capabilities. Observers cheered at the sight of *Columbia*'s ascent. Some, like Frank Gillespie, a retired train conductor from Pennsylvania, remarked on the pride of country the launch inspired. Gillespie said, "It makes you believe in America." What emerged, according to the *New York Times*, was a "collective feeling" by Americans of participating in this new space achievement. The launch and the reactions it elicited gave NASA a tremendous boost of confidence in the shuttle and signified the promise of external support for it: "We did it!" NASA launch controllers at Kennedy exclaimed as *Columbia* left the ground. "Where are the cynics now?"[16] Four days later, a crowd of 250,000 welcomed the ship back to Earth when it landed at Edwards Air Force Base.

The agency believed strongly in the power of seeing the shuttle in flight and wanted as many people as possible to have the experience and be touched by the sublimity of it. Bringing a wide range of people to shuttle launches and landings thus became a major element of NASA's strategy to broaden awareness of and heighten interest in the shuttle as the program continued. The effort built on a tradition begun by public affairs chief Julian Scheer to invite a "broad cross-section of individuals" unaffiliated with the aerospace industry to the Apollo launches.[17]

Given resource and capacity limitations, NASA officials had to be selective about whom they invited to shuttle launches as special guests of the agency. Adopting Scheer's strategy of giving highest priority to "influential members of the general public," they sought people outside of NASA circles whom they believed could, after witnessing a launch, go on to take actions to positively shape the nation's space program. Officials envisioned that these people would be leaders in local governments, labor unions, the media, academia, or the business world. Abrahamson, for one, invited Wall Street financiers to launches and landings to persuade them to underwrite shuttle payloads and infrastructure. The agency aimed to enhance invited visitors' experiences at Kennedy Space Center with special tours and mission briefings. Memoranda circulated to NASA personnel up through the shuttle's final launches solicited recommendations for guests who would fit the profile of influencers. Former NASA official Alan Ladwig observed that those invited to attend launches would "write us nice notes afterward," but the agency generally did not follow through to measure the impact of their guests' experiences.[18]

High on NASA's list for targeting invitations to launches were women and people of racial and ethnic minority groups. Public opinion studies made clear to agency officials that NASA, human space flight, and the shuttle had no constituency among these groups.[19] As Duff expressed it, human space flight "was seen by most women as a white enclave of macho engineers playing with expensive toys, and the perception was that there was not [a] real future for women in this area. . . . The same thing was true of minorities, particularly blacks. In the case of blacks, we didn't even have males on our side. The minorities, especially blacks, thought it was a white man's area." In response to public and political pressure to expand the social diversity of the astronaut corps in the 1970s, NASA had selected some female and Black astronauts; the agency consequently invited people sharing these astronauts' demographics to attend their launches. Longtime NASA staffer Josie Soper recalled that in the shuttle's early days NASA would "approach a different audience of people to come and be there and witness it firsthand, because the philosophy was this is how you're going to bring people on board and how you're going to get support for the program."[20]

At NASA's invitation, hundreds of officials from the National Association for the Advancement of Colored People and the Urban League attended the launch of STS-8, Guion Bluford's first mission. In preparation for Sally Ride's launch in 1983, NASA welcomed women in senior positions in the government and private sector, those with access to the media, and those the agency felt were most likely to have been critics of NASA. According to Duff, some six hundred of NASA's targeted invitees accepted, including feminist activists Gloria Steinem and Betty Friedan. NASA continued to bring women and members of various minority groups to launches of missions with crewmembers with relevant tie-ins.[21]

The agency also saw launch attendance as instrumental to reaching another targeted group: elementary and secondary school students and teachers. Capturing the interest of children who might then attract the support of their parents for the space program was one aspect of why connecting with youth was important to NASA. But the agency also wanted to build the next generation of scientists and engineers who one day would carry on with its work. NASA officials believed that inviting students and educators to shuttle launches and landings would inspire their commitment to the space program and to science and technology generally. The agency's education office would often plan conferences in conjunction with these events to inform them about the mission. Sometimes NASA awarded science fair and other competition winners with trips to see shuttle flights. Further, the agency's education office assisted teachers with integrating the shuttle in classroom curricula to "build a sound educational base for future STS missions, payload activities, and their scientific enhancement."[22]

Just as NASA welcomed the media and prominent members of society to launches to help convey the worthwhileness of the shuttle program to others, so too did agency officials see merit in inviting artists to help tell NASA's story. They believed that creative depictions of shuttle missions would humanize the program and make it more meaningful to citizens. Administrator James Webb had initiated an art program during the Apollo days, but a new program director, Robert Schulman, reinvigorated it in the shuttle era after it had endured a lull correlating with the transition from Apollo to shuttle. The program proved popular among artists, with hundreds contacting NASA each year to participate. Beginning with STS-1, NASA selected a handful of artists to cover each mission's launch and landing, providing them with what Schulman called "access like they've never had before" and giving them special tours of the space center facilities.[23] Compensated with $1,500 each, the artists converted the inspiration they drew from the events they witnessed into works on canvas in a variety of artistic styles, which NASA exhibited at the Smithsonian National Air and Space Museum and in national traveling shows for millions to see. During the 1980s and 1990s, the agency expanded the program to include poets and musicians to help commemorate and convey the meaning of shuttle flights in other expressive, nontechnical ways.[24]

Creating Close-up Mission Experiences

A 1981 NASA publication reflected: "In past exploration, a few hardy souls ventured out, while the rest waited for months or years to hear what they had found. In the Space Age, we are all explorers. Through the miracles of modern communications we have watched together as these new worlds have been revealed."[25] More than ever before, NASA officials strived to make human space flight a shared experience during the shuttle era. Beyond putting the vehicle on display, NASA officials harnessed a broad suite of available communications technologies to make the sights, sounds, and action of operating shuttle missions accessible to as many people as possible.

Equipping the shuttle system to give the public visibility into the program's activities was not a measure to which everyone at NASA readily agreed. The agency demonstrated the ability to use television as both an internal and public communications tool on Apollo 7, but the mission's engineers and astronauts had regarded it as both a cost and safety distraction. Consequently, Scheer had faced difficulty persuading Apollo engineers to include video cameras as a requirement for the monumental Apollo 11 mission. Although he ultimately overcame their objections and proved the inordinate benefit of cameras in attracting media and public attention to NASA achievements, resistant attitudes lasted into the shuttle era. Deputy administrator Hans Mark, for one, worried that providing live coverage of shuttle launches and activities could mar NASA's image in the event of a failure.[26]

At the same time, shuttle program planners recognized the value of a video system for monitoring and analyzing shuttle launches, landings, and mission operations. They established a system of cameras, relay satellites, and transmission lines to support the new vehicle. NASA public affairs staffers eager to share facets of shuttle missions with the public found an ally in Johnson Space Center director Christopher Kraft, who took a broad view of what would define the shuttle's success and agreed to allow them access to the shuttle television system for outreach purposes. The dozens of long-range cameras NASA had installed around its premises relayed, as well as any technology could, the drama of these events for television viewing by millions, and NASA's public affairs office established the means to distribute the feed to private and public broadcasters. High-resolution cameras positioned in the orbiters and on the helmets of spacewalking astronauts did double duty for mission operations and public engagement, giving crewmembers and mission controllers views of hardware in the payload bay while offering public viewers unprecedented looks at the work of astronauts in space. Indeed, NASA public affairs officials saw early on that televising astronauts' activities such as fixing and deploying satellites and conducting scientific experiments in orbit would offer the public ringside seats to the shuttle's accomplishments.[27]

For example, millions watched in amazement as astronauts serviced the Hubble Space Telescope in the 1990s and 2000s.

Public affairs officials and others at NASA who advocated for increased visibility of shuttle missions pressed to upgrade camera resolution and introduce new viewing opportunities. With the 2002 launch of STS-112, the agency mounted cameras on the shuttle's external tank that provided live video of the vehicle's ascent after leaving the launchpad until the jettison of its solid rocket boosters. William Readdy, who oversaw NASA's human space flight program at the time, explained that NASA took the action to demystify and humanize space flight and make it more inclusive. Bob Jacobs, NASA news chief in the early 2000s, maintained that STS-112 was one of the agency's most popular launches in many years because "we did something different and we allowed [people] to experience space flight." However, proponents continually faced opposition from shuttle program engineers who, concerned primarily with missions' technical success, did not appreciate these cameras. Only after the 2003 *Columbia* accident, which occurred because a chunk of foam on the external tank broke off and struck the orbiter's wing during launch, did shuttle program managers recognize the importance of being able to view the outside of the orbiter.[28]

The agency was also committed to relaying the experience of space flight to people in unprecedented ways, often doing so in partnership with outside entities. During the late 1960s, a company that came to be known as IMAX Corporation burst onto the scene boasting a capability to shoot movies using 70-millimeter film for projection on oversized screens. The Mercury, Gemini, and Apollo astronauts had used 16-millimeter cameras to record their mission activities. The agency recognized during the 1970s, however, that the relatively spacious shuttle could accommodate the bulky new cameras and aspired to produce an IMAX film that would make viewers feel like virtual participants in shuttle missions. As proponents within NASA struggled to find the funds and work through the logistics, the IMAX Corporation independently made its own film, *Hail Columbia!*, about the first shuttle mission. Though it did not contain any footage shot aboard the shuttle, the film captured the attention of Beggs and Abrahamson, who viewed the movie when it was

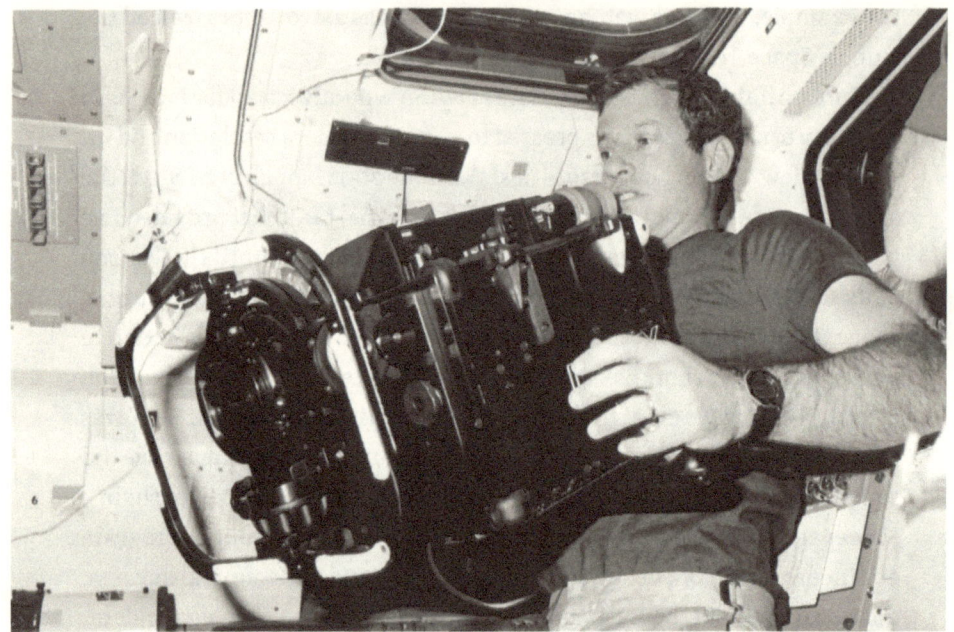

FIGURE 4.3. STS-41C mission specialist Terry J. Hart holds an IMAX camera in the middeck of the space shuttle *Challenger* in 1984. Photograph from NASA.

screened at the National Air and Space Museum. Highly impressed, they, along with astronauts who saw the film, became advocates for flying an IMAX camera aboard the shuttle.[29]

In pursuing IMAX projects aboard the shuttle, NASA was, according to Duff, "true to its philosophy that the people of the world should to the greatest extent share and have a sense of participation in the space program." Because NASA could not give the IMAX Corporation preferential access to shuttle missions over other media outlets, the agency worked out a deal for the museum to make the film with NASA's cooperation. Lockheed Corporation, a major shuttle support contractor, became a corporate sponsor. In a 1983 letter to the museum's acting director, Walter Boyne, Beggs consented to allowing astronauts to be trained to operate the IMAX camera on shuttle missions with the objective "to help create the most effective possible film for communicating the potential and beauty of mankind's newest environment—Outer Space."[30] An IMAX camera flew on four shuttle missions during 1984 and 1985, resulting in *The Dream Is Alive*, a film that depicts many aspects of life aboard the shuttle and premiered at the museum in 1985. The IMAX partnership continued for the next few decades and led to the production of films that captured the Hubble Space Telescope servicing missions, the shuttle's visit to the Russian *Mir* space station, and the construction of the International Space Station (see fig. 4.3).

The agency also formed a partnership with Cinema-360, a consortium of four US planetariums that desired to fly a camera aboard the shuttle to make a film about the vehicle designed for domed screens. Agreed to by Beggs in 1983, the project called for the consortium's 35-millimeter camera to fly aboard up to three missions.[31] The camera flew on two missions, STS-41B and STS-41C, in 1984. When weight limitations arose on its next scheduled flight, NASA worked out an agreement between Cinema-360 and the IMAX Corporation for the IMAX camera to capture footage for the consortium. The completed film, *The Space Shuttle: An American Adventure*, debuted in planetariums in 1985.

NASA also strived to make people feel more closely connected to the shuttle and the astronauts by allowing them to listen in on communications between ground controllers and shuttle mission crews in flight. The agency had provided such access to domestic and foreign media for the first few shuttle flights. NASA partnered with AT&T to enable people to call a designated telephone number on a pay-per-minute basis. The service, dubbed Dial-a-Shuttle, became publicly available with the STS-5 mission in November 1982 (see fig. 4.4). More than one million people dialed the 1-900 number to enjoy the service that AT&T advertised as "just the thing for armchair astronauts." NASA turned to NSI to administer the popular service beginning with the next shuttle mission. Because callers would often dial in when the shuttle was out of range or crews were sleeping, NSI provided live commentary, highlights, and interviews to ensure that callers never encountered dead air. With up to five million people calling in for some missions, NSI and its successor, the National Space Society, continued to offer the service into the 1990s. In 1995, the National Space Society and Rockwell sponsored Dial-a-Shuttle for the first docking of the shuttle *Atlantis* with the Russian *Mir* space station.[32]

NASA used still other techniques to engage the public with mission crews and their activities to relate the experience and significance of

DIAL-A-SHUTTLE®
**THE
ULTIMATE
LONG
DISTANCE
CALL.**

1-900-CALL-NASA
99 cents per minute

Call Dial-A-Shuttle® to listen in on the astronauts and cosmonauts as they dock the space shuttle *Atlantis* and the Russian *Mir* space station for the very first time. Call Dial-A-Shuttle® 24 hours a day beginning on June 1.

Using our special interactive voice response system, you can decide if you want to hear live communications, summary reports with the latest information from the Dial-A-Shuttle® team at the Johnson Space Center, feature reports on the mission, biographies of the astronauts and cosmonauts, or features on the space shuttle and *Mir*, including how Rockwell and Russia's RSC Energia developed and built the joint docking system. You'll also hear mission previews with important information on times such as lift-off, docking and landing, and more!

Sponsored by:

National Space Society
Washington, D.C.

Rockwell

Call 1-900-225-5627. The charge for this call is 99 cents per minute. You must be over 18 or have parental consent to call. The National Space Society is solely responsible for the content of this service.

FIGURE 4.4. An advertisement for the Dial-a-Shuttle initiative in the National Space Society's *Ad Astra* magazine. Courtesy of National Space Society.

space flight. Just as it had during the 1960s, the agency continued to regard its astronauts as one of its most valuable public relations assets. Agency officials had not regarded astronauts through the Apollo era as particularly well-versed in public and media relations, however, and therefore required its first class of shuttle astronauts to take public speaking classes. As former Johnson Space Center director George Abbey explained, this measure allowed the astronauts to "portray NASA the right way." NASA encouraged and often required shuttle crews to go on public speaking circuits following completion of their missions, where they were met with audiences enthusiastic to hear from them. Beggs said that when he toured with STS-1 crewmembers John Young and Robert Crippen, "We wandered around the country talking to people about the experience, and the turnout in almost every city we visited was enormous. . . . The thing that surprised me and made me feel [good] was the fact that we got the general public interested, and they hadn't been all that interested in the interim." In 1983 alone, NASA astronauts supported nearly 2,500 appearances in 43 states and 20 foreign countries, reaching an estimated 2.3 million people.[33]

NASA officials were determined to ensure that astronauts' roles in engaging with the public would also become part and parcel of the shuttle missions themselves. As with the use of cameras for public engagement purposes, not everyone at NASA considered activities intended to meet public communications or education goals to be important or even appropriate activities for astronauts during missions. Shuttle program managers tended to be highly focused on each mission's technical success, the fulfillment of planned scientific objectives, and the completion of tasks for shuttle customers. It was only through the advocacy of public affairs officials that missions beginning with STS-8 included a half-hour press conference featuring the crew in orbit. Shuttle program officials also objected to some initiatives that public affairs staffers championed to publicize missions. When NASA's external relations officials endorsed a proposal by *Sesame Street* in 1985 to fly Big Bird's teddy bear, Radar, on STS-51L, NASA associate administrator for space flight Jesse Moore overruled it. Moore noted that honoring the children's television show's requests for astronauts to appear on its shows was fine, but flying a character, and requiring any involvement of the astronauts with it, "would set a precedent that we would rather not establish." Moore added: "I am becoming increasingly concerned that the flight of such nonessential mission items in the Orbiter can have a negative impact on the NASA image. . . . While I am a major supporter of educational and public affairs initiatives, I believe we can find more appropriate means to deliver our message to the taxpayers."[34]

In still other cases, astronauts were personally interested in engaging with the public during their missions but it took some effort to get shuttle program managers to see the value of these activities. STS-9 astronaut Owen Garriott, an amateur ham radio operator, carried within his personal kit a handheld antenna, which he used to communicate with hams around the world during his off-duty hours. NASA had accepted Garriott's request to carry and use the antenna provided it did not interfere with mission activities or run afoul of safety requirements.[35] His communications were the first between astronauts and people on the ground outside of official NASA channels.

The agency ultimately saw the public relations merit of such engagement. Starting with STS-51F, NASA flew the Shuttle Amateur Radio Experiment (SAREX) to further test ham radio equipment in space as well as to communicate with amateurs and youth. Recognizing SAREX's role in generating "wide-spread public interest" and making young people feel like participants in the shuttle program, the agency partnered with the American Radio Relay League, the Amateur Radio Satellite Corporation, and ham radio operators around the world to fly SAREX on many more shuttle missions, allowing student groups to ask questions directly of shuttle (and eventually space station) crews in orbit. NASA selected school groups to participate based on their completion of an application and proposal. Patricia Palazzolo, a Pennsylvania teacher who spearheaded efforts for her school district to participate in SAREX contacts, reflected that "no one can watch the faces of those students talking via amateur radio to an astronaut and doubt the 'inspiration factor.'" In addition, NASA regularly conducted live in-flight video downlinks with schools for students to ask questions and watch astronauts perform demonstrations in space, reaching more than six million students through these connections. Astronaut

Donald Thomas reflected: "For the first time, we made access to space available to the classroom, and many teachers and students from across the country and around the world were able to participate."[36]

Likewise, it took some time but NASA officials ultimately warmed up to a proposal by Carolyn Sumners, a staff member at the Houston Museum of Natural Science and Johnson Space Center, to have astronauts play with different toys in the microgravity environment of space—with the idea of teaching school-age children about physics. The plan was not originally sanctioned by NASA as an official mission activity, and the STS-51D astronauts who agreed to participate had to carry the toys as part of their personal belongings and run the experiments during their free time. Sumners noted that personnel at NASA Headquarters were concerned that authorizing astronauts to play with toys on missions that cost taxpayers billions of dollars to fund could tarnish NASA's credibility and image as a good steward of federal resources. Agency officials, however, came to see the value of the Toys in Space initiative as a means to interest students in science, math, and engineering and make shuttle missions more relatable, even if the optics were not completely serious. NASA flew toys as part of the official payload on shuttle flights in 1993 and 1996 and later aboard the International Space Station, creating video recordings of the astronauts' experiences for dissemination to schools. A booklet summarizing the project explained, "This toy cargo gave the Space Shuttle one more role in extending human access to the space environment. With the addition of a few pounds of toys, the Shuttle mid-deck became a space classroom where astronauts could teach the nation's children about life in space."[37]

NASA's commitment to making young people feel like virtual participants in the shuttle program went still further. Much of the effort was driven by the Reagan administration, which believed that American science and math education had reached an abysmal state in comparison to that of its worldwide counterparts, threatening the nation's economic and military security. President Ronald Reagan viewed space as an inspirational teaching tool, and in 1984 he introduced an initiative called Operation Liftoff to harness NASA resources in science, engineering, and math education. Syndicated columnist Jack Anderson regarded the

FIGURE 4.5. President Ronald Reagan announces the start of the Young Astronaut Program at a White House ceremony on October 17, 1984. Photograph from Ronald Reagan Presidential Library.

shuttle and human space flight as particularly motivational. He suggested to Reagan that the initiative could include decentralized clubs around the country that youngsters could join, with the goal of immersing them in knowledge about the space program wherever they lived. With Reagan's support, NASA partnered with NSI to develop the Young Astronaut Program "to involve young Americans more directly in our space program" (see fig. 4.5). According to Anderson, who became chairman of the White House–based Young Astronaut Council, the new program would "give the aerospace industry a constituency of millions of kids who believe in space."[38] NASA, NSI, and Reagan envisioned the Young Astronaut Program enrolling children at the elementary and middle school levels in local chapters across the country, where they would learn about the space program through NASA-provided materials and participate in hands-on projects, contests, and sponsored trips related to space. The program ultimately came to be managed out of the White House, with grants provided to chapters by private sector sponsors and NASA serving as a partner in the effort.

The Young Astronaut Program partners went to great lengths to make membership in the organization highly appealing to children, their parents, and teachers. The program blasted chapter application materials to all seventy-seven thousand elementary and middle schools in the United States. To heighten the sense of participation it pledged to fly aboard the shuttle the charter certificates of the first one thousand chapters formed. NASA and the US Postal Service approved a plan by which Young Astronaut Program members could send personal postcards on a shuttle mission. The Young Astronaut Program succeeded in capturing the imaginations of those its organizers targeted: three years after its founding, it boasted 14,000 chapters and 350,000 student members.[39]

Of course, riding aboard the shuttle would provide the ultimate student learning adventure, and in 1985 Anderson suggested to Reagan that a Young Astronaut Program member be permitted to fly on a mission. While NASA administrator Beggs would "not rule out" the possibility one day, the agency kept its focus on giving children opportunities to experience what it would be like to be an astronaut while keeping their feet on the ground.[40] An Alabama state agency had established Space Camp at its US Space and Rocket Center in Huntsville, Alabama, in 1982, and attracted students from across the nation and from other countries to participate in weeklong sessions of astronaut training and learning about the space program. NASA field centers followed suit by providing inspiring opportunities to students in their regions.

Agency personnel got directly involved in some projects. Johnson Space Center education employees helped a local Texas elementary school assemble its own shuttle mockup on its playground to be used for classroom instruction and local Young Astronaut chapter meetings and camps. The US Department of Education also helped the project by providing a $30,000 grant.[41] Employees at NASA's Lewis (now Glenn) Research Center in Cleveland, Ohio, meanwhile, converted school buses into mock shuttle orbiters and outfitted them with computers, scientific laboratory equipment, and two-way radios. They selected elementary and high schools from around the region to participate in simulated shuttle missions, assigning some of the students to roles as astronauts aboard the "orbiters" and others to serve as mission planners and flight controllers, scientific researchers, and public relations specialists. During 1985 and 1987, NASA Lewis, with the support of Rockwell and other private and community sponsors, facilitated the simulated missions, which ranged from a few hours to a full day. Students donned space suits, conducted experiments, and worked through challenges with their peers on the "ground." A report that staff at Lewis compiled on the projects testified to their value not only in motivating students' interest in space, science, and math but also in exciting entire communities about the space program.[42]

NASA shared the shuttle with the American public by giving citizens opportunities to see and connect more closely with the vehicle. These efforts allowed them to suggest that a new, democratic era in space flight had descended upon America. And while the enthusiastic reactions of many citizens to these opportunities indicated that they believed the shuttle belonged to and would benefit them, these approaches to engagement were just one prong of NASA's strategy to bolster the viability of the new space transportation system.

5

Using the Space Truck

NASA had made a bold promise to the world in 1972. In an attempt to gain the support of elected officials, the Department of Defense, the space science community, and the American public, the agency pledged that the shuttle would provide economic, versatile, and routine transportation to space. With its spacious crew cabin and a cargo bay capable of carrying payloads of up to sixty-five thousand pounds, the shuttle would offer plenty of capacity to support civilian government and military needs while making space accessible and relevant to a wide range of new users.

The Nixon administration and Congress chose to fund the shuttle based on NASA's claim that the new vehicle would serve as an all-purpose carrier and Mathematica's analyses indicating that it would prove economical if flown at least a few dozen times per year. While it seemed there would be plenty of NASA and Department of Defense payloads to fill the shuttle's cargo bay, the agency would also be able to recoup operating costs by leasing onboard space for payloads and experiments. By the time NASA's "space truck"—as many within NASA, the aerospace industry, and the media came to refer to the shuttle—started flying in the early

1980s, the Reagan administration suggested that it would be the primary launch system for civil government and national security users. It would also be made available to commercial users and foreign governments.[1] Thus, while NASA strived throughout the 1970s and 1980s to overcome technical and cost challenges associated with the shuttle's development and operations, it also expended considerable effort to solicit a broad range of users. Their active participation as paying clients and innovators was deemed crucial to realizing NASA's new vision for human space flight and ensuring the shuttle's economic and political viability.

But who would want to use the shuttle? Some groups the agency envisioned as potential users of the new space transportation system had not expressed a clear desire for its capabilities. The two groups that NASA officials recognized as most instrumental to the shuttle's success but that were not fully sold on using it included university-based scientists and corporate satellite owners and researchers. NASA worked to create demand for the system among these entities and enroll them as regular users of the vehicle. Finding paying customers for the shuttle was critical, but equally important to realizing the vision of an accessible spacecraft was accommodating new segments of the American, and global, citizenry as users of the vehicle. NASA aimed to broaden the base of shuttle program participants by making excess cargo capacity available to a variety of experimenters at low cost, creating dedicated opportunities for students to participate in shuttle-based research projects, and allowing nonscientific payloads to fly aboard the vehicle. In opening space flight opportunities to new users, the agency's human space flight, science, education, and public affairs offices collaborated and sometimes clashed with one another regarding how to achieve several aims. These included demonstrating the shuttle's utility, remaining accountable to taxpayers, responding to various interests in using the shuttle, providing equal access for a variety of scientific and other purposes, and maintaining the quality of scientific activities conducted aboard the vehicle.

Achieving these objectives required NASA officials to coax some entities to participate as shuttle users while determining the suitability and priority they accorded to groups that solicited NASA, ranging from entrepreneurs to artists. All the while, agency officials had to balance promoting the shuttle with commercial sector concerns about the government-owned vehicle competing with private efforts to develop and offer launch services. Ultimately at stake for NASA was figuring out how to enroll various publics in ways that served them while enabling the shuttle to become the multipurpose vehicle the agency had envisioned. NASA officials strived to ensure that their chosen policies and paths to engage a variety of publics as shuttle users would bestow credibility on the agency and justify continuing human space flight.

Adopting a Customer Orientation

The National Aeronautics and Space Act of 1958 had directed NASA to "arrange for participation by the scientific community in planning scientific measurements and observations to be made through use of aeronautical and space vehicles."[2] Academic scientists in astrophysics, solar physics, Earth and planetary science, and the life sciences had grown accustomed to flying instruments on and analyzing data obtained using sounding rockets, high-altitude balloons, and uninhabited Earth-orbiting platforms and deep space probes. They had far less experience conducting investigations on vehicles occupied by humans, having had only limited participation in Project Gemini, the Apollo lunar missions, Skylab, and the Apollo-Soyuz Test Project.

The scientific community was divided over the merits of human space flight as the shuttle era dawned. Some scientists expressed interest in what possibilities the shuttle could provide for researchers. Others maintained that NASA's human space flight initiatives did not give science enough attention, noting that the agency covered shuttle cost overruns by funneling money from research programs. Recognizing this skepticism, NASA officials and shuttle contractors worked to convince scientists of the shuttle's value and the agency's desire to include their investigations aboard the new vehicle. The shuttle, they said, not only would provide the capability to launch large space telescopes and scientific payloads but also would allow access to space by individual researchers. Compared to flying on a sounding rocket, a shuttle-based experiment could be conducted for many days rather than just a few minutes in the space environment—and at higher altitude. Also, the anticipated frequency of flight would allow for reflights of experiments when necessary. Johnson Space Center director Christopher Kraft explained, "We now know what man can do in space so we are taking the next logical step. We have opened up a new environment and made it accessible for experimentation."[3]

NASA and its contractors strived to capture scientists' interest in using the new vehicle. A Rockwell brochure from the 1970s targeted at university faculty and graduate students stated that the shuttle would bring "the advantages of space flight within reach of every university researcher with an appropriate experimental objective." Experiments flown aboard the shuttle would take far less time for review and integration than what was required to fly payloads on expendable vehicles. The brochure indicated that "universities will become centers of warm support for [the] Shuttle, because it will give them access to the space environment that is otherwise available only to a few investigators in long-term programs." NASA went further by announcing opportunities for researchers to fly aboard the shuttle with their experiments as "payload specialists." The agency suggested that researchers would be able to take flights to "supervise and check" on their experiments due to the shuttle's "easy and routine access to space."[4]

NASA provided many occasions for scientists to participate in shuttle-based research. The Office of Space Science and the Office of Applications released solicitations for academic researchers interested in life sciences and materials sciences to propose experiments with potential societal benefits and commercial applications. NASA's Langley Research Center in Hampton, Virginia, developed the Long Duration Exposure Facility (LDEF), a cylindrical, Earth-orbiting spacecraft the size of a school bus that contained seventy-six trays to hold passive experiments. After releasing a call for proposals in 1976, Langley worked with the Universities Space Research Association to solicit experiments with the aim of encouraging participation by academics with no space research experience. The LDEF launched aboard STS-41C in 1984. It carried 57 competitively selected experiments investigating topics such as materials, thermal systems, electronics and optics, and power and propulsion. The experiments were developed by 200 researchers from 7 NASA centers, 9 Department of Defense labs, 8 foreign nations, 21 universities, and 33 private companies.[5] Although NASA planned to retrieve the LDEF one year later, scheduling issues and then the 1986 *Challenger* accident delayed its return until 1990.

Perhaps the most significant shuttle-based opportunity NASA created for researchers was through the Spacelab module and payload carriers. Designed to provide scientists "continuing, economical access to space," Spacelab was developed at NASA's invitation by the European Space Research Organization (ESRO), representing nine Western European nations. In addition to valuing the shuttle for its ability to increase access to space for Americans, President Richard Nixon had expressed keen interest in international cooperation in NASA's space initiatives following Apollo. Dale Myers, NASA associate administrator for manned space flight, recommended to NASA administrator James Fletcher that Europe could be involved in the shuttle program without being in its critical path by fabricating a laboratory that could be installed in the shuttle's cargo bay on selected missions.[6] Envisioned by NASA to fly twice per year, the Spacelab provided a pressurized module for researchers interested in solar physics, astrophysics, materials processing, life sciences, and Earth observations to work in a "shirtsleeves" environment. It also included external pallets for experiments requiring direct exposure to space. In February 1977 NASA's Office of Space Science and ESRO's successor, the ESA (European Space Agency), selected experiments proposed by 222 scientists from around the world to fly on the first Spacelab mission. Soon thereafter, the researchers selected from among themselves two payload specialists to operate the experiments on the shuttle on their behalf. After Spacelab 1 flew on STS-9 in November 1983, NASA and the European Space Agency flew more than twenty Spacelab missions.

Attracting university space scientists was important for NASA to prove the shuttle's utility, and the agency financially supported much of their research. But NASA also needed to realize the commitment it had made to its political stakeholders and the American people to transform the shuttle program into a self-sustaining operation. Making good on this promise meant the agency would need to find enough customers willing to pay to leverage the shuttle's spacious cargo bay, mission duration, and other attributes.

In some respects, this seemed like a natural transition for NASA: space enthusiasts within and outside of NASA had already articulated a future in which people would exploit space for a variety of purposes. However, marketing the human space flight program to generate a user base and serve public clients represented a major cultural shift for the agency. Some at NASA thought that becoming a space access service was not befitting of an exceptional engineering organization. Shuttle operations director Chester Lee hoped the shuttle would become financially self-sustaining quickly so that NASA could turn over operations to the private sector, shed its image as "a trucking company," and get back to the "sexy" work of exploring space.[7]

NASA's top officials during the 1970s and early 1980s did what they felt was necessary to abide by the new economic and political realities and embraced a conceptualization of the shuttle as a business enterprise for the foreseeable future. They embraced the idea that NASA's new human space flight program would not only focus on meeting its own objectives but also offer reimbursable services to others. Fletcher attested to the shuttle's contributions to commerce in various speeches. Robert Frosch, who assumed the agency's helm during the Carter administration, thought of NASA's responsibility in human space flight as "finding out things for other people and building the technology to enable us to keep doing this."[8] Frosch's successor, James Beggs, acknowledged in retrospect that NASA's ability to generate sufficient payload demand and fly frequently enough to cover its operational costs was "a pipe dream." Nonetheless, he was determined to uphold the vision of the shuttle as an all-purpose carrier, pledging to the Reagan administration: "Okay, you got me. That's the way I'll sell it."[9]

Beggs's commitment was sustained by President Ronald Reagan's desire to develop commercial uses of the shuttle and space. Reagan saw the shuttle as integral to his political theme of renewed American confidence and pride. He noted in his 1984 State of the Union address: "Just as the oceans opened up a new world for clipper ships and Yankee traders, space holds enormous potential for commerce today."[10] The fact that the European Space Agency, led by France, had also developed its own rocket, Ariane, and was marketing it for commercial satellite launches heightened the impetus for NASA to build confidence in the shuttle's performance among prospective payload owners and to make the spacecraft competitive.

NASA's Office of Space Flight, the organization within NASA Headquarters responsible for the shuttle's development and operations, configured itself for business in the mid-1970s. The office's head, John Yardley, a McDonnell Douglas executive during the Mercury, Gemini, and Apollo programs, established a customer service branch and appointed Lee as its leader. Yardley also invited Jon Michael Smith, a marketing manager from Computer Sciences Corporation, to join NASA and work with Lee to help promote the shuttle. The team quickly got to work developing strategies to attract customers for the shuttle.

According to Smith, Fletcher and other NASA senior leaders encouraged "out-of-the-box thinking" and expected sophisticated approaches to marketing. Smith consulted with George Abrams, who had promoted the denture cleanser Polident, to review NASA's shuttle marketing plans. NASA awarded the Battelle Memorial Institute two contracts to advise on pricing policies as well as sales tactics to use and avoid. A three-year study conducted for NASA by General Electric and released in 1976 indicated an untapped market of companies outside of aerospace that could harness the microgravity environment for research, new product development, and manufacturing in fields ranging from pharmaceuticals to metallurgy. The study also recommended that NASA share information about the shuttle with nonaerospace companies early and frequently and address their concerns about handling of proprietary and patent issues.[11]

Smith, Lee, and other agency officials embraced the advice. They conducted seminars about the shuttle for prospective users at professional conferences and trade shows worldwide. Associate administrator for space flight James Abrahamson made quarterly trips to Wall Street to educate bankers on NASA's plans for, and success with, the shuttle to entice them to invest in shuttle-related commercial endeavors. While US government regulations prohibited NASA from spending funds to woo clients with fancy meals and parties as the European quasi-government launch company Arianespace could, agency officials strived to persuade audiences that the shuttle would offer a robust, well-priced platform for conducting business. They touted the shuttle's flexibility, capabilities, and expected ability to fly with airline-like regularity. Frosch instruct-ed NASA representatives to avoid employing a "razzle-dazzle style" at industry meetings but instead to have direct, private interactions with potential customers. Indeed, the agency was determined to present an image of a responsive and cost-effective operation.[12]

NASA and its shuttle contractors also worked to drum up business for the vehicle by emphasizing their commitment to customer service. An agency brochure from the early 1980s, *NASA: Ready to Serve You in Space*, promoted the shuttle's capabilities along with the agency's dedica-tion to customers. McDonnell Douglas, the Spacelab payload integration contractor and developer of a system to propel satellites into higher orbits from the shuttle, took out a *Wall Street Journal* ad stating: "If you want to carry payloads into space, come see McDonnell Douglas. . . . We have the people, the technology, and systems, and the experience to get you there" (see fig. 5.1). Rockwell, the orbiter's prime contractor, sponsored a 1978 ad depicting the vehicle's open cargo bay inscribed with the words "Rock-well can fit you in" and detailing the many ways the company could assist shuttle customers. One Rockwell brochure stated plainly: "We want to help you become a user of the STS," while another offered to help shuttle users with payload design, fabrication, integration, and other planning activities through Rockwell's STS utilization planning service.[13]

NASA took many measures to build the confidence of prospective sat-ellite deployment customers. After the new space transportation system began flying, crews of some of the earliest missions began displaying signs emblazoned with "We deliver!" upon successful deployment or retrieval of satellites (see fig. 5.2). The agency embraced the motto, producing a 1983 publication titled *We Deliver* that boasted, "Twenty-five years of hands-on experience assures you of the most reliable, flexible, and cost-effective launch system in the world. . . . You can't beat manned reliability. . . . In all the world, you won't find Shuttle's equal."[14] The agency published *We*

▶ FIGURE 5.1. Space shuttle contractors took out newspaper and magazine advertisements, such as this one from payload integrator McDonnell Douglas, to encourage satellite owners, researchers, and manufacturers to use the shuttle. Reproduced by permission of Boeing Company.

Room for Rent.

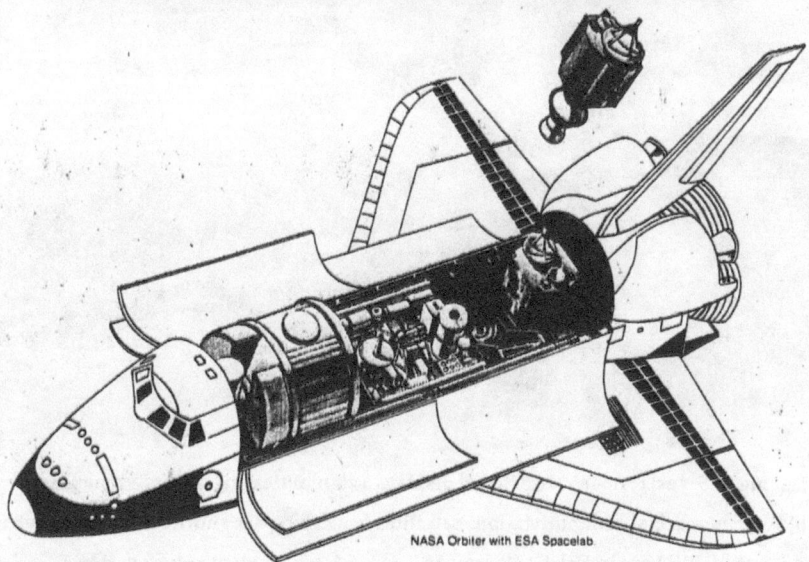

NASA Orbiter with ESA Spacelab

If you want to carry payloads into space, come see McDonnell Douglas.

Until now, few scientists and industries could tap the weightless environment of space for investigative research and manufacturing. Cost was just too great. But Shuttle is changing all that.

NASA has named McDonnell Douglas payload integration contractor for Spacelab flights on Shuttle.

We have also commercially developed a propulsion system called PAM for payloads which require boost to higher orbit from the Shuttle. This means, on a space-available basis, we can put your satellites, research and manufacturing projects onto the Shuttle, and with

our Payload Assist Modules, into space, where you want them.

What do you want to do? Grow ultra-pure crystals? Develop new life-saving drugs? Conduct materials research? The possibilities are as limitless as space itself.

If space holds promise for you, contact McDonnell Douglas now. We have the people, the technology, the systems, and the experience to get you there. Contact: M. J. Schmitt, A3-110, McDonnell Douglas Astronautics Company, Huntington Beach, CA 92647. Telephone: (714) 896-5043.

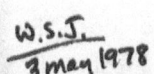

W.S.J.
3 may 1978

MCDONNELL DOUGLAS

FIGURE 5.2. During STS-5 astronaut Vance Brand displays a sign indicating the crew's successful deployment of two commercial communications satellites from the space shuttle *Columbia* in 1982. Clockwise from Brand: Bill Lenoir, Robert Overmyer, and Joe Allen. Photograph from NASA.

Deliver in other languages to market the shuttle at the 1983 Paris Air Show, where NASA representatives eager to best the French Ariane vehicle were equipped with computers to provide prospective clients with on-the-spot projections of shuttle availability.[15] NASA's 1984 shuttle marketing plan indicated that the agency was determined to be viewed by customers as "capable of solving their problems, reducing their technical risks, and providing new opportunities for space-based ventures."[16]

Lee's customer service office, meanwhile, developed pricing policies and a range of incentives to broaden NASA's shuttle customer base. In 1977 NASA established reimbursement policies for satellites to be carried in the orbiter's cargo bay. The agency gave priority to defense or its own major payloads but otherwise assigned customers to missions on a first-come, first-served basis once they provided a down payment of $100,000. The customer service office established a standard price of $18 million plus a facility and equipment use fee of $4.3 million for dedicated use of the shuttle's cargo bay and payload deployment. NASA arrived at the price based on its requirement to recover the shuttle's estimated $9.3 billion cost over its first twelve years of operations, during which it expected to fly 560 missions. The agency anticipated adjusting the prices after some time based on actual costs and mission demand but kept them at this level initially to make the shuttle competitive with the Ariane, which carried a $25 million price tag. NASA offered prorated prices for customers whose payloads would fill only part of the cargo bay and could be paired with others. It also gave discounts to clients willing to fly their payloads on a standby basis. Customers could pay additional fees for priority scheduling, reflight guarantees in case of a problem with a payload, and optional services such as retrieval of payloads from orbit, data processing, and use of upper stages to boost payloads to orbits the shuttle could not reach. The agency also aimed to give satellite owners what Abrahamson called "an extra level of confidence" by offering them the prospect of flying their own representatives as payload specialists aboard the shuttle along with their payloads.[17]

The Office of Space Flight also wanted to interest companies in using the shuttle to conduct experiments with potential commercial applications. Agency officials had claimed that vaccines, enzymes, protein crystals, and other substances could be produced in purer forms in microgravity than was possible on Earth. Enrolling manufacturers of these products as shuttle users would further the new vehicle's economic posture and validate its value for developing products and services that would benefit society. In 1979 the Office of Space Flight created a concept called a joint endeavor agreement, through which NASA would share with customer companies the costs and risks associated with their use of the shuttle for manufacturing research. By signing a joint endeavor agreement with NASA, a company could develop an experiment intended for commercial purposes using its own funds and NASA would provide it space on a shuttle mission at no cost. Companies would retain the results of the research. NASA anticipated that the companies would capitalize on the proprietary data and commercialize the technology at their own expense and purchase space on future shuttle flights to manufacture commercial products.[18] The agency extended its incentive to fly corporate payload specialists to these research flights as well.

NASA declared the shuttle operational and open for business after four test flights. Even before STS-1 flew, the agency had manifested (put payloads on its schedule) nearly one hundred satellite payloads, 22 percent of which were owned by commercial and foreign clients. Agency officials boasted to Congress and other audiences that the shuttle's cargo bay was fully scheduled with major payloads into 1985. The agency flew the shuttle's first two commercial satellites, Satellite Business System's SBS-3 and Telesat Canada's *Anik C-3*, on STS-5 in November 1982. The communications industry in the United States and abroad proved to be the agency's primary non-US-government customer, making up 110 of 130 commercial and foreign payload reservations by 1984.[19] By the time of the fateful 1986 *Challenger* flight, NASA had deployed more than two dozen satellites owned by US commercial companies and foreign entities. The agency also proved the ability to use the shuttle to retrieve satellites for in-space repair and redeployment by astronaut crews with the Solar Maximum Mission spacecraft in April 1984 during STS-41C. In November of that year, shuttle crews retrieved and relaunched two customer satellites, Indonesia's *Palapa-B2* and Western Union's *Westar-VI*, which had not reached proper orbits during an earlier shuttle mission because the kick motors that were intended to position them had failed.

Several companies and university researchers signed on to conduct experiments aboard the shuttle with the belief that they might evolve into profitable enterprises. McDonnell Douglas teamed with Ortho Pharmaceuticals Company, a Johnson & Johnson subsidiary, to sign the first joint endeavor agreement with NASA in 1980. They wanted to fly a series of experiments to test whether they could perfect a process called electrophoresis to manufacture highly purified drugs that could treat diseases such as diabetes and anemia. The technique had been investigated in Earth laboratories, but studies during Apollo, Apollo-Soyuz Test Project, and Skylab missions suggested that employing the technique in microgravity could enhance the purity of hormones, enzymes, and other medical products. McDonnell Douglas wanted to refine the process for large-scale production aboard future shuttle missions and make it commercially viable. Charles Walker, a McDonnell Douglas engineer, became the shuttle's first corporate payload specialist, accompanying the experiment three times. He explained, "When the Space Shuttle flew, we were just all enamored with the prospect of ready access to space for all interests in this nation. . . . [T]he prospects that we all believed were going to flow from that were just going to be enormous."[20]

The list of commercial experimenters planning to use the shuttle went on. A company with no aerospace affiliation, GTI Corporation, signed the second joint endeavor agreement with NASA for four flights of a furnace to understand how metal alloys behave in weightlessness. If it worked, GTI planned to pay NASA to fly the furnace on subsequent flights and lease it to parties to use for their own research. NASA also signed an agreement with 3M to fly three experiments in 1984 and 1985 to grow crystals of organic compounds in space and make thin films from them; the company went on to contract with NASA for seventy-two flights over a decade. A Texas-based startup company called Microgravity Research Associates booked a slot on a 1986 shuttle mission to try to produce gallium arsenide integrated circuit chips in microgravity as an alternative to silicon for electronics equipment. Lehigh University and NASA's Marshall Space Flight Center teamed up to fly an experiment in middeck lockers on five shuttle missions to develop small, uniformly sized latex spheres for uses in calibrating scientific instruments and drug research. The US Bureau of Standards certified the material in 1984 and planned to begin marketing it as a standard reference material in mid-1985.[21]

Companies sometimes approached NASA about alternative forms of involvement with the shuttle, and NASA officials accommodated them where collaboration would advance the vision of a multipurpose spacecraft. The agency signed an agreement with Spacehab, a company formed in 1983, to fly modules in the payload bay filled with additional middeck-sized lockers for experiments. The Coca-Cola Company proposed flight testing a Coke-filled dispenser specially designed for drinking carbonated beverages in microgravity. The company hoped the experiment would give them insight on how taste preferences can change but also aspired to secure contracts for the container's use by NASA or private space ventures. After the company publicized its intent to work with NASA, PepsiCo approached the agency about flying a prototype can containing its own soft drink. Determined not to show a preference for either beverage, NASA reached agreements with both companies in 1985 and flew their cans on the same mission, STS-51F.[22]

While seeking to make the shuttle commercially viable, NASA established certain limits to partnering with external entities. Officials were unwilling to enter agreements that they felt could compromise taxpayer trust in the agency. One such proposal NASA rejected was for commercial operations of the shuttle program. NASA had long contemplated turning over shuttle operations to a commercial entity, freeing the agency to focus on advancing technology and giving more work to industry. William Sword, a Wall Street investor, offered to purchase an additional orbiter for NASA in exchange for marketing rights and profits from the entire fleet, which then stood at four. Building a fifth orbiter to increase the shuttle program's capability had been the subject of ongoing discussion among NASA, the Department of Defense, the Reagan administration, and Congress. Negotiations with Sword failed, however, because the investor's desire for NASA to fund the orbiter's operations costs didn't sit well with Abrahamson and other NASA leaders.[23] NASA achieved a middle ground later when it consolidated contracts with major aerospace companies to manage shuttle operations.

Ultimately, NASA found attracting steady commercial use of the shuttle challenging. While the agency enjoyed a full manifest for major payloads through 1985, much of the crowding stemmed from the shuttle debut's delay from 1979 to 1981. The slate of commercial payloads for the latter part of the 1980s was not nearly as dense. Some satellite owners expressed concerns about the shuttle's schedule reliability when they were bumped for Department of Defense payloads and when flight preparations proved too complex for the agency to maintain the number of annual missions it had projected to achieve. Some companies made reservations on both the shuttle and Ariane to safeguard their flight opportunities, and NASA promised some customers launches on expendable US launch vehicles if the shuttle proved to be unavailable when needed. When the agency announced that beginning in 1985 it would increase its prices by 85 percent, to $38 million for use of the full payload bay, to recover higher-than-anticipated operating costs, the shuttle promised to become all the less competitive with Ariane.[24]

The Reagan administration also presented challenges to the shuttle's commercial success. While President Reagan praised the shuttle, his admiration of the vehicle stood in tension with his commitment to support the private launch sector. Even charging commercial and foreign payload owners $38 million would not fully cover launch costs, and NASA's subsidy irked those in the White House, Congress, the Department of Transportation, and the private sector who advocated for the development of a US commercial launch industry. The agency argued for raising the price to $71 million to cover its "out-of-pocket" costs for flying commercial and foreign payloads. Meanwhile, the administration directed NASA to commit to begin recovering the full costs of shuttle use from customers by 1988 and to auction off access to the vehicle if demand surpassed availability.[25]

The agency also struggled to sustain private research ventures aboard the shuttle after a few years. Companies raised concerns about the costs of using the shuttle versus Earth-based facilities for manufacturing purposes. Some also considered the long time frame involved from making an investment decision to acquiring data on a shuttle experiment, as well as the advance of new ground-based laboratory breakthroughs and research patent rights. Johnson & Johnson left the electrophoresis venture with McDonnell Douglas to pursue similar research in Earth-based laboratories after another company identified a way to develop proteins using genetic engineering breakthroughs.[26]

Beggs recognized the obstacles. While a big market was unlikely to materialize in the short term, he remained confident that the shuttle would help NASA develop commercial applications over time: "I figured that sooner or later somebody in the industrial world was going to figure out something—maybe in the solid-state business, maybe in the metallurgical business, maybe in some arcane business where somebody wants a strange, wonderful product in nanotechnology—that requires that kind of environment."[27] That optimism carried NASA as it persisted in promoting opportunities for external entities to become actively involved with the shuttle and making the space transportation system a viable, accessible, and relevant enterprise.

Offering Low-Cost Opportunities for Experimenters

NASA's plans called for each shuttle flight to carry in the orbiter's cargo bay one or more primary payload, such as a Spacelab module or pallet or a satellite owned by NASA, the Department of Defense, or a commercial entity. As Office of Space Flight officials began assigning these payloads to shuttle missions in the mid-1970s, they found that the cargo would not completely fill the shuttle's available volume or sixty-five-thousand-pound weight capacity. They examined alternatives for what to do with the empty space, concluding that in some cases the shuttle would need to fly with ballast in the cargo bay to balance out the primary payloads.

Yardley and Smith recognized, however, that rather than flying containers of sand, NASA had an opportunity to maximize the system's utility while broadening its customer base: the shuttle could carry small secondary payloads. They devised the Small Self-Contained Payload program, in which NASA would fill spare room in the shuttle's cargo bay with canisters purchased by private entities interested in flying their own space experiments. Yardley and Smith believed that the opportunity would appeal to small companies and university researchers who had ideas for manufacturing products in the space environment: they could conduct experiments at relatively low cost and risk in the canisters and then reserve more shuttle space later for full-scale production.[28] Dubbing the program the Get Away Special (GAS)—a riff on TWA's promotional airfare to Hawaii at the time—they began offering the canisters at prices well below cost, ranging from $3,000 to $10,000 depending on the size and weight of a payload a customer desired to fly (see fig. 5.3). Yardley announced NASA's plans for the program at the annual meeting of the International Astronautical Federation in Anaheim, California, in October 1976.

The initiative was revolutionary on several levels. The GAS program represented yet another change in the way the agency did business. It was driven by one's willingness to pay, rather than by the typical competitive processes and peer review NASA used to select scientific projects. A down payment of $500 was all that was needed to hold one's place in the launch queue, which was determined on a first-come, first-served basis. NASA

figure 5.3. Get Away Special canister payloads installed in space shuttle *Discovery*'s cargo bay await the launch of STS-91 in 1998. Photograph from NASA.

also did not require purchasers to share their results with the agency as they did of scientists who received grants for research projects. And as with joint endeavor agreements, the agency would not seek rights to users' data, inventions, or patents unless the results were likely to have a significant public health or safety impact. This meant that GAS cans afforded access to space for what Abrahamson called a "new class of customers"—and in doing so embodied the democratic vision for the shuttle NASA had promulgated. A NASA publication reflecting on the program's achievements stated that, in addition to being of use to "professional" experimenters, the GAS program "was an avenue, never before available, to space experimentation for the 'man and woman on the street.'" The *Washington Star* declared that with the GAS program, "for the first time since the space age began 20 years ago, the average citizen is being given a chance to participate in the program with something more tangible than tax dollars."[29]

NASA's Office of Space Flight personnel believed that exciting new uses of space would be realized by opening the GAS program to the broadest community possible. That community, as they imagined it, encompassed entrepreneurs, civic groups, academic institutions, and companies in various industries in the United States and worldwide. A September 1977 brochure noted, "Any responsible person, organization, or institution can take advantage of this program." The key requirement for participation was that the purchaser's proposed use had to address a scientific or technological objective. Payloads could not jeopardize the orbiter or crew safety or interfere with other payloads. Any experiments involving live animals would require approval by the Life Sciences Division of NASA's Office of Space Science, which followed the United States Department of Health, Education, and Welfare guidance for laboratory animal use. Alan Ladwig, who served as the GAS program manager at NASA Headquarters in the mid-1980s, recalled: "This wasn't just work in your backyard and go and throw it down in the cargo bay. You had to go through a series of safety reviews just like anything else did. But . . . it opened the aperture for more than just NASA-funded scientists to do things."[30]

While the Office of Space Flight's marketing personnel had to work hard to secure major payload customers, the GAS program seemed to sell itself. The first GAS customer provided the agency with a check for his $500 earnest money the day after Yardley announced the program. R. Gilbert Moore, a Thiokol executive from Utah with a passion for science education, noted that "since the earliest days of the space program there has been no way to get students involved. Space was the province of the intellectual and physical elite. But NASA, in what I call its infinite wisdom, has come up with a way of getting around all that." Moore purchased a GAS can for his two sons to fly an experiment to find out whether the family's pet lizard would regenerate its tail in space. He donated the remaining room in the canister to Utah State University for student experiments. The ten experiments packed into the container became the first to fly in a GAS canister, launched aboard STS-4 in 1982. Moore, together with Utah State professor Rex Megill, purchased additional GAS containers for use by students at other universities and a Utah public school district.

The two men began a program at Utah State for high school students to propose experiments and fly them in the canisters. Dubbed Project Enterprise, the endeavor earned the praise of Senator Ted Stevens of Alaska, who called it "an outstanding example of public participation in support of space experimentation."[31] Moore went on to purchase several more GAS containers throughout the program's existence.

NASA received hundreds of letters of interest and down payments for canisters. Just a year after announcing the GAS opportunity, the agency had accepted earnest money from eighty-five parties from the United States and abroad.[32] They represented universities and high schools, corporations interested in science and engineering research, hospitals, publishing companies, banks, and real estate agencies. They also included individual private citizens and families who wanted to experience the thrill of flying their own payloads.[33] Domestic and foreign research institutes, many of which had never been involved with space research, purchased and flew GAS cans to investigate how space could enhance biological, material science, and other research. Two people, Ellery Kurtz and Howard Wishnow, flew four of Kurtz's oil paintings and other art materials as a GAS payload on STS-61C in 1986 to understand the effects of space flight on art conservation. George Park Jr., assistant vice president of the Park Seed Company, flew a GAS canister on STS-6 full of forty varieties of fruit and vegetable seeds to understand how to ship seeds into space and produce a viable crop aboard a space station. A longtime space enthusiast, the seed purveyor indicated that he also pursued the opportunity because "by doing this I can maybe make the space program more real. Bring it down on a day-to-day level. Down to the soil, if you will, for the people in America. This is our program."[34]

The number and sorts of people who became involved in the development of GAS payloads exceeded the Office of Space Flight's original expectations. The Tokyo newspaper *Asahi Shimbun* reserved a GAS can and conducted a contest for readers to propose uses, choosing from among seventeen thousand ideas an experiment suggested by two Japanese high school students to synthesize pure artificial snow crystals.[35] The canister flew on STS-6. A planetarium in Mississippi purchased a container for a movie camera to record views of Earth from space for incorporation into an educational film. A connection between an RCA executive and the principal at an underprivileged New Jersey high school allowed a group of students to fly an experiment on STS-7 in 1983 to study microgravity's effects on the behavior of carpenter ants. Although the ants were dead upon return to Earth (and probably never reached space alive due to prelaunch mishandling of the GAS canister), some three hundred students participated in the project as experiment designers, fundraisers, and publicists. Movie producer Steven Spielberg donated a GAS container to Caltech, while aerospace company TRW donated a canister to the Explorer Boy Scouts. A Texas beer distributor reserved three GAS cans for use by the University of Texas–El Paso, two school districts, and the bordering cities of El Paso and Juarez, Mexico.[36]

The Office of Space Flight recognized that many GAS customers had little, if any, experience with space experiment development and took measures to boost their likelihood of success. Although officials decided early on against assisting participants in formulating experiments or interpreting data, the office did offer technical advice. Yardley solicited the help of the Sounding Rocket Division at NASA's Goddard Space Flight Center to work with GAS customers and to design and develop the containers. Eventually renamed the Special Payloads Division to reflect a broader suite of responsibilities, the Goddard office developed a GAS experimenter handbook and conducted symposia to convene program participants to discuss their experiences and results. As the program grew, the agency proposed a regulation that aimed to ensure equitable allocation of GAS canister flight opportunities to government, commercial, and educational customers. Industry users had been purchasing more than twice the number of GAS canisters as educational institutions and governmental entities combined. The approved rule modified the first-come, first-served policy to allow for a rotational sequence among the three types of users. To address the growing backlog of GAS containers, NASA contracted with Teledyne Brown Engineering to develop a "bridge" to fit across the shuttle's payload bay and accommodate twelve canisters. The bridge first flew with a full suite of a dozen canisters on STS-61C in January 1986.[37]

Philosophical differences about the program sometimes arose between the Office of Space Flight and other NASA offices. Noel Hinners, NASA associate administrator for space science, argued that experiments "paid for by the local 'Lions Club'" and without participation by a "reputable scientist" were "wasteful and pointless and should be discouraged." Shuttle operations director Lee took exception, saying that the figure of merit for GAS cans was not scientific rigor but instead the educational process and the introduction of young people to the space frontier. Several operating concerns also cropped up. The GAS program had been designed to require no additional services like electrical power or payload deployment from canisters, but Lee's office sought to maximize the capabilities available to customers. He told shuttle payload integration and development managers at Johnson Space Center that he believed NASA's accommodations for small payloads on the shuttle should "strike a balance between customer needs and the difficulty of satisfying them."[38]

Only against what Lee described as "a great deal of opposition within NASA" did his office get approval to provide customers with standardized containers and the services of an astronaut to operate on-off controls for experiments during their missions.[39] The Office of Space Flight also was able to increase the sophistication of the GAS containers to include features such as windows and opening lids. But it faced particularly stiff resistance from shuttle safety officers when it came to allowing customers to eject small payloads or substances from GAS canisters. Two artists affiliated with MIT wanted to fly a container housing a high-powered electron gun that would project a beam of electrons into Earth's ionosphere to create an artificial aurora. They dubbed the project New Wave Ruby Falls after an underground waterfall in Tennessee. The conflict was ultimately resolved in favor of providing this capability when Yardley's successor, Abrahamson, got involved. As one who viewed the shuttle as "a means of making the dreams of Americans come true," Abrahamson's decision led NASA to fly and eject small experimental satellites from GAS containers for the first time on STS-51B in 1985.[40]

For researchers who wanted still more capabilities for their experiments, NASA offered alternative carriers. The Hitchhiker and its derivative, Hitchhiker Junior, provided flights in the shuttle's cargo bay for payloads too large for a GAS container but too small to be affixed to a Spacelab pallet. The first Hitchhiker launched carrying three experiments on

STS-61C. As these opportunities had price tags on the order of $1 million, GAS remained a more attractive option to many experimenters. By the eve of the *Challenger* accident in January 1986, NASA had flown 53 GAS payloads on 14 of 24 shuttle missions conducted to date, while 458 reservations remained in the queue.[41] Clearly, the agency had sparked many people's enthusiasm for participating as users of the shuttle, and it worked to satisfy the demand.

Providing Access to Student Experimenters

From its earliest days NASA aimed to connect with the nation's elementary, secondary, and postsecondary education communities to build the space agency's future workforce. As the shuttle era rolled around, NASA officials sought to inspire student interest in the space program and realize the shuttle's democratic vision by providing opportunities for young people to interact with astronaut crews in orbit and to participate in simulated missions. Some education proponents in and outside NASA believed the agency could go further by allowing students to get involved in shuttle-based research. Thus, in addition to the GAS program, NASA established opportunities to enroll students as active participants in shuttle missions. Doing so required champions for these endeavors to persist in their efforts, as they encountered opposition from colleagues who doubted the wisdom of flying student experiments.

NASA had started inviting proposals from academic scientists and graduate students in the 1960s to participate in space research projects using a variety of platforms, including human-tended spacecraft, free-flying probes, sounding rockets, and high-altitude balloons. Space-based research experiences for less advanced students, meanwhile, were quite limited. Undergraduate students could participate in cooperative education programs and summer internships hosted by some NASA field centers. Hands-on space research opportunities for high school students did not exist—until agency officials entertained the possibility beginning with its first space station, Skylab.

In the early 1970s, NASA sought to promote the value of Skylab, an Earth-orbiting laboratory built from spare Apollo components to perform biomedical, environmental, and solar physics research. As part of this effort, NASA's Education Programs Division (as it was called by 1971) and Skylab program officials at Marshall Space Flight Center seized upon a suggestion by a Martin Marietta Corporation contractor to solicit student experiments for Skylab and devised a plan for a competition among high schoolers. Myers endorsed the contest as "an unparalleled opportunity for direct public involvement in the values and benefits of the space program," and Fletcher approved it in August 1971.[42] The following spring, 25 winners were selected from among 3,409 proposals submitted through the competition, which was administered by the National Science Teachers Association. Although the contest concluded late in Skylab's development phase, NASA and the winning students moved quickly through experiment reviews and preparations, and the agency flew nineteen of the projects on Skylab.

Myers had noted in advocating the Skylab student program that it presented a new and valuable way for NASA to connect with citizens. He further said, "We recognize that it represents a commitment to deliver the benefits of the space program to the public in a way that is different from what we are accustomed to, and that it would be a commitment not easily withdrawn." Indeed, the Skylab student program had attracted support from teachers, students, and members of the general public who sent donations to the agency to defray the costs of student projects. The positive reception inspired some within NASA to call for a similar program once the shuttle began operations. As it was, the new vehicle's projected frequent flights meant that NASA could provide more opportunities for student participation without being tied to the tight timelines required to prepare Skylab experiments.[43]

One month after President Nixon approved the shuttle, and while the Skylab student competition was under way, Frederick Tuttle, head of the education division, raised the idea to Charles Donlan, the acting director of the shuttle program at NASA Headquarters. A few years later, Marshall Space Flight Center, which had managed the Skylab student program, contracted with the University of Alabama in Huntsville to examine potential educational uses of the shuttle. The resulting report suggested that NASA could build on the Skylab experience and consider flying high school as well as college student experiments aboard the shuttle. In June 1976, a Skylab student experimenter who was serving as a co-op student at Marshall drafted a concept for a shuttle student experiment program, noting the benefits for NASA, students, and educators alike. James Murphy, a Marshall official, sent the proposal to Philip Culbertson in the Office of Space Flight at NASA Headquarters, encouraging Culbertson to start the program during the shuttle's orbital flight tests as an activity "to which the general public could relate." At about the same time, Frank Hansing, director of university affairs at NASA Headquarters, shared the idea with NASA's third-highest-ranking official, associate administrator John Naugle.[44]

Interest in a student experiment program came from outside the agency as well. Operations Research, Inc., of Silver Spring, Maryland, for example, proposed a program to involve high school juniors and seniors and college students from the Washington, DC, area in research projects aboard the shuttle's Spacelab module. Martin Marietta shared its own ideas for shuttle-era education programs. But perhaps the most influential entity was the Forum for the Advancement of Students in Science and Technology (FASST). An affiliate of the American Association for the Advancement of Science, FASST was a national network of individuals and organizations that believed college students should actively participate in all aspects of science, including the policy process, research, and applications. In 1973 the organization, by then renamed the Federation of Americans Supporting Science and Technology, began to press NASA to allow student experiments to fly on the shuttle.[45]

FASST members offered to help NASA create awareness of shuttle opportunities within the university sector, and the agency accepted, looking to the organization to provide student ideas for experiments that could be flown in space. FASST also compiled an advisory board of prominent and powerful people, including Utah senator Frank Moss, who served on the Aeronautical and Space Sciences Committee. In August 1976 Moss sent a letter to Fletcher suggesting that NASA develop a shuttle student participation program. With the idea already circulating within the agency, Fletcher replied that he was considering a competitive program open to high school and college students.[46]

By February 1977, NASA's Office of External Relations, under which the education and university affairs offices then resided, submitted a detailed plan for a shuttle student program to Fletcher. The initiative would "tap the ingenuity and- resourcefulness of young minds in devising meaningful experiments and demonstrations that will add to our store of scientific knowledge" and "create a spirit of involvement in the resumption of human space flight among students, faculty and parents with a multiplier effect that can impact favorably upon public attitudes toward the space program."[47] But when Fletcher announced his resignation a month later and shuttle-related technical and financial challenges saddled NASA's next administrator, Frosch, momentum behind the initiative slowed to a crawl.

FASST refused to let the issue fade. Members undertook a letter-writing campaign to Frosch. Having collected more than five hundred ideas for experiments, the organization called on NASA to provide students with opportunities to fly experiments in space.[48] FASST also reached out again to influential members of Congress, prompting Adlai Stevenson and former astronaut Harrison Schmitt of the Senate Subcommittee on Science, Technology, and Space to send Frosch a letter stating that students should be considered part of the "user community" for the shuttle. The Senate report accompanying NASA's 1979 authorization bill directed NASA to develop options for expanding student involvement.[49]

With this impetus, the student experiment initiative moved a step closer to becoming reality. Despite the challenges with the shuttle program, the new administrator truly believed that including students was an opportunity to maximize creativity in space research. Frosch said that he did not see the involvement of students in NASA's work as an "add-on or a public relations gimmick, or a let's-do-this-and-see-what-we-can-get-in-the-way-of-a-constituency" effort. Rather, he saw student contests for research as a means "not only to interest new people in space research and in the Shuttle, but as a way of finding people with new ideas and new things to do."[50] By mid-1978, Frosch hired Glen Wilson, a former Senate Aeronautical and Space Sciences Committee staffer, as his assistant for student activities, including development of a shuttle student experiment program.

Wilson set to work on several open issues. He had to figure out, for one, how to finance the program. Budget pressure brought on by shuttle development and other NASA needs meant that the agency had limited resources to administer the program and fly the winners' experiments. Wilson also had to determine who could participate and how the program would be managed. Advocates inside and outside of NASA desired a program available to both high school and college students. While NASA had experience with administering the Skylab high school program, figuring out how to manage the selection process at the college level proved challenging. Wilson worked with FASST to conduct a workshop to help the agency determine the size, scope, and implementation of a college-level program.[51]

Another major area of concern swirled around the scientific value of the program. In contrast with Frosch, Wilson was convinced that the primary purpose was "humanizing" the space program for the public and building a supportive constituency for NASA among students, teachers, and parents. Science was secondary. "You would hope for but not necessarily expect to get good science," he told the steering committee supporting the program's development. Even so, Wilson wanted to make sure that the program stimulated students' interest in science and space and maximized the merit of their work. During the Skylab program the agency did not provide close supervision or require students to report their results, and scientists within NASA were already critical and skeptical. Charles Pellerin of the Office of Space Science at NASA Headquarters recognized student programs as having value as "publicity- and interest-generating devices" but minimal scientific productivity. Richard Young, acting deputy director of life sciences, held an even more pessimistic view: "Students sponsored by a civic organization with the casual participation of a local high school teacher are not likely to learn much and I would guess NASA will often end up in an embarrassing public position."[52]

NASA scientists also expressed concerns about the resources the agency would accord to the program, questioning how it could justify supporting student experiments when it could not afford to fund all good proposals from "qualified scientists." As the shuttle's readiness date continued to slip, NASA's science offices argued that students should not have priority over professional researchers whose experiments faced postponement. Glynn Lunney, acting associate administrator for space transportation operations, raised the concern that putting student experiments in middeck lockers of the crew cabin, as Wilson envisioned, would be unfair to the GAS program's paying customers who had access only to the shuttle's cargo bay.[53]

The program structure that Wilson and others at NASA formulated attempted to reconcile these issues. They designed what became known as the Shuttle Student Involvement Program (SSIP) as an annual nation-wide competition for students to propose experiments that could fit in a shuttle middeck locker and require up to one hour of an astronaut's time during a mission.[54] NASA focused the competition on students in grades 9–12, never obtaining the funding to add a college component. The agency contracted with the National Science Teachers Association to run the competition, which involved selecting up to twenty semifinalists across ten geographic regions and choosing up to ten national winners among them. NASA would pair each winner with a NASA scientist or engineer to help prepare the experiment for flight and would identify an industry or other non-NASA sponsor who could defray participant expenses and provide technical advice. The agency did not guarantee that all experiments would fly, noting they would do so only based on their readiness and the availability of room on a mission.[55]

Beginning with the 1980–1981 academic year, just ahead of the launch of STS-1, NASA ran its first SSIP competition. The agency received more than 70,000 requests for information and 1,500 proposals, from which the National Science Teachers Association's teams of teachers, scientists, and engineers selected 191 semifinalists and 10 national winners. The following year, the agency received 2,800 proposals—nearly double the previous year's total—and chose 20 national winners. NASA selected 27 more winners during the next three competitions. The winning experiments covered a wide range of scientific topics. Eighteen-year-old Todd Nelson of Minnesota proposed to study the motion of moths and bees in microgravity (see fig. 5.4). Daniel Weber of New York sought to understand the effects of weightlessness on arthritis. Seventeen-year-old Karla Hauersperger of North Carolina wanted to examine how space travel affected chromium levels in the human body. NASA received commitments to sponsor the students from the aerospace industry and sectors aligned with the students' experiments, including biotechnology and

FIGURE 5.4. Shuttle Student Involvement Program finalist Todd Nelson of Minnesota (*center*) shows the container for his experiment on the behavior of bees in microgravity to STS-3 astronauts Jack Lousma (*left*) and Gordon Fullerton after a preflight press conference. Photograph from NASA.

pharmaceuticals. Participating students, teachers, and corporate sponsors reported that they held highly favorable impressions of the program and had learned much about the scientific method through their experiences. Many regional and national winners went on to study science and engineering in college.[56]

Even given the years of preparation, transforming the students' ideas into flight-ready projects proved challenging at many points. Resistance to a project to study biofeedback processes came from the astronaut office and an external scientist who aspired to conduct a similar experiment. Further, with an active antivivisection campaign directed at animal research at NASA's Ames Research Center, life sciences division director Gerald Soffen raised concerns about the "public outcry" and "serious mischief to NASA" that could result from flying two student experiments involving rats. One involved breaking rats' bones to study fractures and healing in microgravity; the other would involve injecting rats with an arthritis-inducing serum and dissecting the animals back on Earth. Echoing scientists' earlier criticisms of student experiments, he also noted that the experiments constituted "poor science"—despite the fact that one of the students had received assistance from Pfizer and they both complied with federal and state animal research laws.[57]

Despite these objections, Abrahamson believed in SSIP's importance and transferred the program to his office. He pressed Johnson Space Center to fly SSIP experiments, noting that students "deserve" the opportunity.[58] As a result, Nelson's insect study flew on STS-3 and seventeen more followed before the January 1986 launch of STS-51L/*Challenger*. Many other experiments were in preparation at the time of the accident, and three rode on the *Challenger* mission. President Reagan's State of the Union address, which had been scheduled for the day of the *Challenger* flight and consequently was deferred by one week, illustrated how significant student participation had become to the shuttle program. Tying the work of one SSIP participant to the concept of the American dream, Reagan stated: "We see the dream coming true in the spirit of discovery of Richard Cavoli. All his life he's been enthralled by the mysteries of medicine. And, Richard, we know that the experiment that you began in high school was launched and lost last week, yet your dream lives. And as long as it's real, work of noble note will yet be done."[59]

As SSIP became established for students to participate in shuttle-based research, Bill Kinard of NASA's Langley Research Center pondered another opportunity to engage youth with science enabled by the shuttle. In 1983 Kinard was serving as chief scientist for the LDEF. After the LDEF endured several launch delays, some companies that had planned to fly experiments on the platform withdrew, creating availability for additional experiments. Kinard recognized the educational value of student participation in the SSIP and wished to provide the LDEF's spare room to projects involving schoolchildren. Rather than solicit student proposals, a process that likely could not be completed before the LDEF's launch date, Kinard sought to involve youth in a NASA-led experiment. At the time, the Park Seed Company was preparing to fly plant seeds in a GAS canister

FIGURE 5.5. A packet of space-exposed seed prepared for distribution to schools participating in the SEEDS project. Image from NASA.

on STS-6. Kinard approached the company and convinced Park Seed to fly a followup experiment on the LDEF to test the effects of long-term space exposure on a variety of plant seeds. He discussed with Park Seed assistant vice president Park and director for research Jim Alston having the seeds from the LDEF distributed to students to grow.[60]

The three men agreed to pursue the project, believing it would broaden participation in space research and stimulate interest in science. Dubbed SEEDS: Space Exposed Experiment Developed for Students, the initiative kicked off with Park Seed furnishing 12.5 million tomato seeds, which were placed into five aluminum canisters aboard the LDEF (see fig. 5.5).[61] NASA, for its part, funded, tested, and flew the experiment's hardware and developed promotional and instructional materials for the project. Under a cooperative agreement with NASA for education support services, Oklahoma State University aided the agency in interacting with prospective and participating schools.

Like SSIP, the SEEDS project encountered skepticism about its scientific value. Some scientists affiliated with NASA questioned how the agency would ensure the project's data quality when thousands of schools would grow and measure plants from returned seeds. Some university researchers objected to the fact that they could not receive seeds for study. SEEDS endured far less resistance than SSIP, however, because the space on LDEF would have otherwise gone unused. In addition, NASA made clear that the project's primary aim was education and outreach, not world-class science.[62]

Ultimately, the project enjoyed positive reception across NASA and the education community. Launched in 1984 aboard STS-41C, the SEEDS project remained in space for five years—longer than anticipated because the 1986 *Challenger* accident delayed STS-32, LDEF's return mission. NASA sent some 132,000 experimental kits to more than 40,000 schools in 31 different countries, serving more than 3 million students. Elementary school, high school, and university students alike participated in the project. Each participating entity received at least 50 space-exposed seeds and 50 control seeds. Students grew the seeds and sent reports of their observations to Oklahoma State, which compiled the results for NASA. The project proved to be a successful forerunner of collaborative "citizen science" projects and, equally importantly for NASA, garnered enormous levels of enthusiasm among participants. NASA's publication *SEEDS: A Celebration of Science* shared the scientific results along with dozens of testimonials from teachers, students, and parents explaining how the project benefited them and expressing their gratitude to NASA for the opportunity to participate.[63]

Setting Policy for Nonscientific Uses of the Shuttle

While NASA made strides to give the public opportunities to conduct scientific research and development activities in the years leading up to the first shuttle flight, individuals and small companies also queried the agency about gaining access to the shuttle for other purposes. Some wanted to fly a personally meaningful item or something of symbolic or cultural value. A doctor from New York expressed interest in deploying a small satellite from a GAS can in honor of his son's wedding. Several artists sent checks to NASA to fly works of art aboard the shuttle, and a group of children asked about deploying a small sculpture of a child from the payload bay. Schools and companies inquired about sending mementos bearing their insignias to display after flight. Others had commercial purposes in mind. Some wanted to fill canisters with T-shirts, coins, or other collectible items for resale. A Dallas retailer hoped to fly gerbils in a GAS can with the same intent.[64] People asked about flying human ashes in space to memorialize loved ones. Meanwhile, companies sought to use the shuttle and related ground-based assets to advertise products.

NASA officials in the Office of Space Flight, in collaboration with others throughout the agency, set to work to navigate this new terrain and mediate the democratic imaginary it had established for the shuttle. On the one hand, the Office of Space Flight wanted to sign up as many shuttle users as possible. The revenues that NASA would receive from these additional customers, especially advertisers, could help position the shuttle as an economically sensible enterprise and reduce costs borne by taxpayers. Vice President George H. W. Bush and Senator William Proxmire agreed, also believing that using the shuttle to make the space frontier accessible to as many people as possible would strengthen the nation's presence in space.[65]

On the other hand, some NASA space flight officials worried that allowing people to fly objects they intended for resale or which others might find objectionable could embarrass the agency or mar its credibility. NASA still reeled from the criticism unleashed by members of Congress and the media when, in the early 1970s, Apollo 15 astronauts were accused of selling postage stamps flown on their mission for personal gain.[66] Now the agency had been entrusted to spend billions of taxpayer dollars on a new human space flight initiative, and officials were determined that the pursuit would be reputable. As the proposed uses of the shuttle came to light, some scholars and journalists chastised NASA for not maintaining a "serious" image, while others followed the most eccentric ideas with amusement.[67] Thus, in addition to requiring payloads to comply with specific crew, spacecraft, and live animal subject safety regulations, the agency would insist on some degree of decorum for using the shuttle.

Consequently, NASA made choices about the types of items it would and wouldn't fly for members of the public. In 1978 the agency welcomed shuttle customers, companies, schools, and foreign governments to fly patches, medallions, and other small items meaningful to them aboard a shuttle mission. The items would be compiled in an Official Flight Kit that would be stowed in the orbiter's middeck and returned to the owners after flight. None of the mementos could be flown for economic gain, all had to earn the approval of NASA's administrator, and Johnson Space Center shuttle program management reserved the right to deny items due to size, weight, or any other reason.[68]

The agency also set policies for use of GAS cans. The Space Transportation System Users Service Council, comprising representatives from several NASA offices, evaluated each payload proposal's suitability. While liberally interpreting what constituted a research and development project, they prohibited GAS cans' use for "crassly commercial purposes."[69] Thus, Lee returned Joseph Roberto's $500 for a canister to fly human ashes on the shuttle. Roberto had founded a company called Astro Burial Ltd. with the intent of ejecting a person's cremated remains into space for $2,995. Despite Roberto's petition that his proposal was an experiment in the humanities and would open opportunities for modestly sized businesses in the space program, Yardley rejected the appeal, explaining: "Flying human beings alive or deceased on the Shuttle for the purpose of fulfilling an individual desire to orbit the Earth does not meet the requirements at this time for an acceptable payload."[70]

NASA officials also set standards for advertisers who sought to connect their products with the shuttle. Companies had considered affiliation with the space program to be a sign of prestige since NASA's earliest years. Ads highlighting the Gemini astronauts' consumption of Tang were perhaps the most notable example. By the start of the 1980s, companies with no business ties to NASA were releasing ads mentioning the shuttle to highlight their products' quality. A Volkswagen ad in the *New York Times* referred to its "Vanagon" model van as a "space shuttle," showing an image of the van's roomy interior. *Playboy* took out a full-page ad in the *Times* that displayed the shuttle orbiter as a symbol of American prosperity in an attempt to entice companies to advertise in the men's magazine.

As long as ads did not use NASA's name or insignia or otherwise suggest endorsement, the agency generally would not object, appreciating the positive publicity from ads that identified NASA with excellence or that reinforced the shuttle's prominent role in society. NASA's public affairs officers also permitted companies promoting many types of products— from cars to fashion to orange juice—to conduct film and photo shoots at NASA's visitor centers.[71] However, they drew the line at advertising that would involve use of the shuttle or related operational assets. For example, NASA nixed a proposal by Los Angeles marketing executive Robert Lorsch to sell plaques for $1 million apiece on which companies could inscribe "supportive, non-commercial" statements. Lorsch wanted the agency to mount the plaques in the shuttle's crew compartment and focus a television camera on each one for sixty seconds while in flight. After taking a fee of nearly 18 percent, Lorsch would give the remaining funds to NASA in a sponsorship model he likened to that used by the Olympics.[72]

The Office of Space Flight's responses to the rejected proposals commonly suggested that NASA might review the possibility of flying unorthodox payloads in the future. Terence Finn of NASA's legislative affairs office responded to inquiries from the public and Congress by stating that NASA might develop policies for "unconventional" and "innovative and imaginative ideas for the use of space" as shuttle operations matured. Lee's letter to would-be celestial undertaker Roberto stated that the agency was cognizant of the importance of "extending the involvement in space activities to a wider range of human interests." Lee indicated that NASA might consider "payloads based solely on individual desires" once shuttle flights became routine. He noted that the agency was reviewing the possibility of flying non-research-and-development-related payloads in the shuttle's cargo bay under NASA's regular commercial rate. Starting at $1.8 million, however, the price tag would be prohibitive for small businesses, so other approaches would be necessary for them to engage.[73]

Lee's marketing advisor, Smith, recognized that the shuttle was "bigger" than science. The various public inquiries suggested to Smith and others at NASA that the agency had to find a way to allow people of all backgrounds to gain access to the shuttle to "do their thing, whatever it was." This sort of engagement would honor the shuttle imaginary and, many at NASA hoped, help to enroll people as supporters of NASA's activities. As Tony Maull, a NASA official who got involved in evaluating nonscientific proposals, explained, "What motivates the American people to support the space program—and it is their program, not a hobby-shop for us—isn't just for the reasons of science and engineering, but perhaps as important, for the reasons of poetry, the spirit of adventure."[74]

Office of Space Flight personnel informed NASA's deputy administrator, Alan Lovelace, of the brewing public interest in using the shuttle for nonscientific purposes. In 1979 Lovelace established a committee to make recommendations. He appointed NASA chief scientist Frank McDonald to head the committee and called for it to include "expertise in the non-technical fields from which subject proposals eminate [sic]." McDonald and others within NASA who believed that such an initiative would trivialize the space program expressed their opposition to a nonscientific payload program for the shuttle. Ultimately, however, Lovelace authorized the Office of Space Flight to draft a policy.[75]

Preoccupied with juggling existing payload customers, the Office of Space Flight deferred the task until 1984. That May, NASA proposed in the *Federal Register* to amend its rule on shuttle use to "broaden the range of potential launch service customers and increase the access of the general public to the [shuttle]." Published in final form in August, the rule declared NASA's willingness to consider flying, on a space-available basis, "cargoes that do not meet the definitions of national defense payloads; communications, weather, or other high-technology satellites; materials sciences/processing payloads scientific experiments; engineering test articles or other similarly technical cargoes routinely considered for flight as conventional or self-contained payloads." The agency would create an internal Nonscientific Payload Evaluation Committee to meet quarterly to determine the propriety of flying any payload that otherwise met NASA's safety and operating standards. Untethered satellites and payloads containing human or animal life would not be permitted, and the agency reserved the right to reject proposals focused on gaining publicity. Finally, the policy stated that NASA would employ no more than 10 percent of the shuttle's flight capability for nonscientific payloads and required customers to reimburse NASA for operations costs of nearly $1,500 per pound plus integration fees.[76]

In 1985, the Nonscientific Payload Evaluation Committee scoured more than two hundred requests it had received to fly payloads through the Nonscientific Payload Program. The vast majority failed to fully satisfy the policy's criteria. But after consulting with the National Endowment for the Arts, the committee decided that a work by Massachusetts College of Art professor Lowry Burgess fit NASA's requirements. "Out of all proposals made to us, only Lowry Burgess' seemed to be innocent," Maull said, referring to the stipulation prohibiting flight of nonscientific payloads for blatantly promotional purposes. Burgess had been one of many artists who had queried NASA in the late 1970s about sending artwork into space. While he had then envisioned flying a piece for astronauts to deploy outside of the shuttle, the policy's requirement that no payload could be left in orbit prompted him to reconceive his plan. He proposed instead a piece he named *The Boundless Cubic Lunar Aperture*. The five-inch cube of bronze-tinted glass contained water Burgess had collected from eighteen rivers and eighteen glaciers, wells, geysers, and ponds, inside of which floated a smaller, empty glass cube. Burgess's intent was to place the artwork, once returned to Earth, in a petrified tree within a rock formation near Walden Pond in Massachusetts. For Burgess, who had struggled to comprehend the United States' role in Vietnam and his own life's purpose, putting his artwork in space provided a means "to express through art the scientific observation of order and harmony in the universe."[77]

Burgess' six-pound sculpture was assigned to fly aboard a spring 1986 shuttle mission as NASA's first nonscientific payload, although the flight hiatus following the *Challenger* catastrophe in January of that year delayed the artwork's launch until 1989. The selection of this nontraditional payload epitomized the agency's budding commitment to redefining relationships with the public and making the shuttle broadly accessible. Creating pathways for the public to engage as shuttle users was hardly straightforward, but NASA made significant strides within the first few years of shuttle operations to open space and build user communities from many walks of life in ways that had only recently been fantasy.

6

Creating Space for New Flyers

NASA's introduction of a broad range of users and uses of the shuttle was a key way of making the vehicle meaningful to more Americans and legitimizing resources spent on human space flight. But efforts to engage diverse segments of society substantively in the shuttle program did not stop there. Since NASA's earliest days, the focus of its human space flight programs was on the astronauts. Thus, agency officials recognized that who they involved as flyers aboard the shuttle would be crucial in determining how connected to human space flight the nation's people would feel.

The astronauts throughout the 1960s were cut from a similar mold: they were white male aviators, mostly drawn from the ranks of military test pilots. In 1958 NASA had considered conducting an open competition to recruit explorers of extreme environments—mountain climbers, spelunkers, and other adventuresome types—to serve as its first human space flyers. Ultimately, however, President Dwight Eisenhower directed the agency to tap military aviators because they had already undergone rigorous physical and psychological preparation. NASA announced its first class of Mercury astronauts in April 1959, to include three pilots

FIGURE 6.1. Many Americans followed with excitement the pioneering flights of the Mercury astronauts but their homogeneous backgrounds as white male military test pilots made it more difficult for some segments of the US population to relate to human space activity. Clockwise from top left: Alan Shepard, Virgil I. ("Gus") Grissom, Gordon Cooper, Scott Carpenter, John Glenn, Deke Slayton, and Wally Schirra. Photograph from NASA.

hailing from the air force, two from the navy, and one from the marine corps (see fig. 6.1). These seven men, and those selected over the next few years to support the Gemini and Apollo programs, became the most visible symbols of the space agency and its achievements as the nation sought to convey technological and ideological superiority to the Soviet Union during the 1960s. NASA, in partnership with *Life* magazine and other news outlets, fostered a public image of the astronauts as heroic, chaste, and invincible—even when their personal lives sometimes belied these attributes—in an effort to project an infallible aura around the US space program.[1]

Indeed, NASA set a precedent that its human space travelers would remain an invaluable element of its public outreach efforts. But the demographic makeup of the astronaut corps became increasingly out of sync with the national focus on promoting workplace diversity and opportunities for all. Thus, as part of the shift to a democratic imaginary for human space flight, NASA emphasized that the shuttle would accommodate a wide variety of people on board. With a winged, plane-like design that would reduce launch and reentry stresses on astronauts, the new space transportation system would be accessible to individuals with different credentials and who were not necessarily in the same physical condition as the military test pilots who had dominated NASA's astronaut corps for a decade.

As was the case with expanding users and uses of the shuttle, NASA personnel were divided on inviting new sorts of astronauts. Many at NASA Headquarters regarded introducing opportunities for different types of flyers as a necessary means to bolster the human space flight program's relevance. Some astronauts and others at Johnson Space Center, however, worried about the potential impact on the cost and safety of shuttle missions as well as the flight priority of career astronauts already waiting for mission assignments. At stake on both sides of the issue was the agency's credibility and legitimacy. Proponents saw broader access and democratization as a way to boost the agency's public standing, while those opposed believed that the agency could best preserve its reputation by remaining elite. Although these two perspectives remained in tension with one another, the quest to make the shuttle a routinely operating, accessible vehicle ultimately pushed the agency toward broadening the range of shuttle flyers.

Expanding Professional and Social Diversity in the Astronaut Corps

At Eisenhower's direction, NASA focused its selection of the Mercury astronauts in 1959 on military pilots. More specifically, the agency sought men who were younger than forty, in impeccable physical condition, and had college engineering degrees and extensive experience test flying fighter jets. Although the agency's recruitment for Gemini and Apollo astronauts in 1962 expanded eligibility to civilians, it still required the same education pedigree and emphasized a background piloting high-performance aircraft. These origins gave way to what Tom Wolfe calls a "right stuff" astronaut culture within NASA—one in which a man would "go up in a hurtling piece of machinery and put his hide on the line and have the moxie, the reflexes, the experience, the coolness" to get his spacecraft under control.[2]

Where the shuttle was concerned, however, piloting would be only part of what was needed to operate the spacecraft. While pilot astronauts would focus on flying the system safely and providing overall authority onboard, another cadre of space flyers would be required to manage the shuttle's systems and the scientific and technical payloads on each flight. The agency would hone a new breed of career astronaut: the scientist, engineer, or physician trained not as a pilot but as a "mission specialist."

Even during the "right stuff" heyday, many scientists advocated that NASA consider expanding the astronaut ranks to include expertise from their profession. Although the one-man Mercury flights were restricted to proving the operational capabilities of capsules, Gemini and Apollo flew multiple astronauts, spent longer in orbit, and required fewer systems tests, thus allowing time for scientific research. In the early 1960s, the Space Science Board, many members of which were involved in planning scientific observations and research for the Apollo flights to the moon, recommended that scientifically trained astronauts be included in the lunar missions as crewmembers.[3]

NASA heeded the recommendation and in early 1964 made plans to include a "scientist-astronaut" on each Apollo lunar landing flight crew. The Office of Manned Space Flight and the Office of Space Science and Applications at NASA Headquarters worked with the Manned Spacecraft Center and the National Academy of Sciences to develop selection criteria. Candidates would need to possess a doctorate or equivalent experience in a science, be less than thirty-four years old, and—like pilot astronauts— have 20/20 uncorrected vision and be able to pass a class I military flight status physical. More than one thousand people applied that October. NASA sent four hundred dossiers to a special committee of the National Academy of Sciences for review and received recommendations for sixteen nominees. From these, NASA chose six: one geologist, two physicians, and three physicists. In 1966 the agency undertook a second round of scientist-astronaut selections and, using a similar process, identified eleven from another pool of nearly one thousand applicants.[4]

Though the Manned Spacecraft Center participated in shaping the scientist-astronaut selection criteria, mission managers and astronauts were unenthusiastic about the scientists' arrival. With the number of scheduled Apollo missions countable on two hands, the Astronaut Office's chief, Donald Kent "Deke" Slayton, told the new recruits not to expect flights.[5] Ultimately, however, at the insistence of NASA deputy administrator George Low, who championed the value of including scientists, mission planners assigned geologist Harrison Schmitt, one of the original scientist-astronauts, to Apollo 17. Another three of the original scientist-astronauts—Joseph Kerwin, Owen Garriott, and Edward Gibson—each flew on one of the three Skylab research missions launched in 1973. But many years would pass before the shuttle was ready to fly and NASA could make mission assignments for its other scientist-astronauts. Officials found interim assignments for some to support various programs and functions around the agency, but several resigned out of frustration when future opportunities looked bleak.[6]

As shuttle program planning got into swing, NASA officials established a subcommittee of its Space Program Advisory Council to look again at the value of scientist-astronauts within NASA's cadre of career space flyers. The subcommittee, chaired by NASA chief scientist Homer

Newell, advised not just continuing but expanding the agency's engagement of scientists as astronauts "in consonance with space shuttle science and applications needs." The reviewers further opined that NASA should organize to "reflect appreciation and approval of science and its support" and recommended introducing the position of a scientifically credentialed mission specialist on shuttle missions to liaise between payload and flight operations. NASA acting associate administrator John Naugle, also a scientist, agreed with the subcommittee's assessment, stating: "Your study has shown how the scientist astronaut, bridging the sometimes wide gap between scientific and flight operations points of view, can contribute to a productive Space Shuttle science program." Consequently, the agency adopted plans to fly at least one mission specialist on each shuttle flight.[7]

Given the array of payloads that NASA officials anticipated flying, they envisioned mission specialists coming from the life or physical sciences, medicine, or engineering but operating onboard as generalists. These career NASA astronauts would need to gain familiarity with the shuttle's operating systems but would not be required to go through pilot training as their Apollo-era predecessors did. The agency would also relax the medical standards for these astronauts, as they would not have direct responsibility for mission safety. In contrast with the original scientist-astronaut standards, the shuttle mission specialists could be of any age and would need to be able to pass a NASA class II, not class I, space flight physical and have 20/100 visual acuity correctable to 20/20 in each eye. NASA director for life sciences David Winter stated, "Our aim is to get the best qualified scientists that we can into space and bring them back safely. . . . Therefore, our approach must be to broaden the medical criteria as widely as we safely can."[8]

In 1976 Christopher Kraft transferred the agency's ten remaining Apollo-era scientist-astronauts to a new Office of Mission Specialists. Later that year, NASA issued its first call for astronauts in more than a decade, noting that the agency sought to add at least fifteen new mission specialists to its ranks. No longer involving the National Academy of Sciences, NASA's recruitment and selection process was driven by agency managers, astronauts, and discipline specialists. They chose twenty mission specialist candidates representing a wide range of scientific and technical expertise from civilian as well as defense organizations. NASA announced them, along with fifteen newly selected pilots, in January 1978. Kraft wrote to Naugle that it was his desire "to give the mission specialists visibility in the Shuttle Program" and to keep them "competent and content professionally." While the new mission specialists' hectic training schedules ultimately precluded from keeping up with their research careers, NASA delivered on giving them active roles in the shuttle program. Upon successful culmination of a two-year training and evaluation period (reduced to one year by 1983), the candidates were converted to career status and assigned to specific missions.[9]

The agency was ready to fly its first two mission specialists in 1983 on STS-5, the flight immediately following four orbital tests of the shuttle system. Both of them, Joe Allen and Bill Lenoir, had joined NASA's ranks as scientist-astronauts in the 1960s, and they were given responsibility to deploy the first commercial satellites from the shuttle's payload bay. In June 1983, STS-7 carried three mission specialists drawn from NASA's 1978 astronaut class. The mission specialist had arrived as a new breed of space flyer that would be part of every astronaut selection cycle and shuttle mission crew throughout the program's operational life.

NASA's 1978 astronaut class also reflected another palpable shift in its approach to making selections. The thirty-five new recruits included six women, three African Americans, and one Japanese American. It was a stark contrast from NASA's entirely white male astronaut corps from the previous two decades. The civil and women's rights movements had led members of the public, civic groups, and Congress to question the lack of racial and gender diversity in NASA's workforce, including the astronaut corps. Why had the agency not selected a female astronaut or a Black astronaut? The Soviets had flown Valentina Tereshkova and brought her safely back to Earth in June 1963. And thirteen American female pilots had, under a private program partly funded by one of the women's husbands in the early 1960s, passed the same physiological tests that NASA had administered to male candidates.[10]

Misogynistic attitudes ran strong among many NASA officials and political stakeholders in the agency's earliest days. According to Brian Duff, who headed NASA's public affairs office in the 1980s, "it took a long time to get women in the position of trust in the astronaut program, unless they were nurses, dieticians, or in other traditional roles." Beyond such positions, only Frances "Poppy" Northcutt was closely connected to human space flight in the 1960s, serving as NASA's sole female flight controller in mission control during the Apollo program. Duff continued: "A number of the old NASA people fought that movement and predicted all sorts of bad things would happen if women flew." Apollo 8 astronaut Frank Borman, for one, stated that women "would have caused more problems than they would have been worth" on early flights, and New Mexico senator Clinton Anderson believed that NASA needed to learn more about space flight "before we hazard a woman in space." In 1970, NASA physician Charles Berry asserted that having women join men on long space trips might be "more comfortable" and "more normal," but more research would be needed to draw a conclusion on the matter.[11]

For the most part, NASA personnel justified the astronaut corps's composition in the 1960s as being necessary to do the job of beating the Soviets to the moon. They argued that they had established specific criteria for astronauts—namely, engineering educations and extensive experience with high-performance aircraft—but no woman or applicant from a racial minority group had come forward with those qualifications.[12] "There is no discrimination whatsoever against candidates of any race, color, or creed," NASA public affairs assistant administrator George Simpson wrote in a 1962 letter to Joan Duge of Williamsville, New York. "If a qualified Negro candidate volunteers for service as an astronaut, you may be sure that he will receive equal consideration with other candidates." Likewise, NASA's Julian Scheer responded to a query from Virginia Allan, president of the National Federation of Business and Professional Women's Club, stating that "NASA has no policy of keeping women out of the astronaut program. However no woman has as yet met all the stringent qualifications which NASA has established for astronaut trainees. . . . NASA has erected no barriers against women in space; we are instead trying to establish the most effective program possible."[13] The agency denied claims of discrimination by Jerrie Cobb, one of the thirteen female pilots who had proven able to pass the agency's medical tests, maintaining that it needed people with test operations experience, which all the women lacked.[14] Yet with so few women and people of color graduating with science and engineering college degrees and as military test pilots, it was no wonder why the few women and Black Americans who did apply during that era failed to break in: the standards *were* the barriers, and NASA seemed unwilling to take them down.

By the early 1970s, NASA could no longer explain away the absence of women and minority groups in its astronaut corps. The agency had reached the moon. While billions of people around the world had marveled at the achievement, plans to complete all scheduled lunar landing missions were on shaky ground as the Vietnam conflict and social issues such as urban decay and poverty demanded attention and resources. Meanwhile, the women's and civil rights movements had moved forward with vigor. The Apollo program increasingly seemed like an anachronism, with the uniform whiteness and military image of the astronaut corps suggesting that NASA was out of step with the times.

Many journalists and pundits vocalized this concern. Joseph Morgenstern raised the issue days before the Apollo 11 launch in a *Newsweek* piece questioning the lunar program's value: "Are the astronauts really representative of humankind and the dreams of humankind, when they are white, male, and American? . . . Do they instead connote a sense of domination, as 'cowboys in a new uniform'?" A 1974 *Los Angeles Times* article by J. K. Obatala titled "We Need to Correct a Space Age Injustice: America Can Still Have a Black Astronaut" suggested that the zeitgeist necessitated a change in NASA's selection process. "Mars and the moon are as much a part of America's future as are Madison Ave., Hollywood Blvd., and the Mississippi Delta," Obatala argued. With everyone footing the bill for the space program, "NASA has an obligation to include all segments of society. . . . From now on, it is essential that the standard of social equality is carried aloft whenever—and wherever—America may explore." Science fiction writer Isaac Asimov, meanwhile, argued in the *Boston Globe* that NASA should "call off" the space program if the agency was unwilling to accept women as astronauts, whom he regarded as "biologically sounder" and "more resistant to stress and less subject to a variety of metabolic diseases" than men.[15]

The agency also faced legal pressures to change. The civil rights and women's movements had prompted the creation of policies and legislative measures that were intended to ensure that citizens received equal treatment on many fronts, including in federal government hiring practices. Like other federal agencies, NASA responded to the directives for affirmative action, taking steps to increase the involvement of minority-owned businesses in procurements and inviting racially diverse university participation in NASA research opportunities. The agency established an Equal Employment Opportunity (EEO) office in 1972 following the passage that year of the EEO amendment to the 1964 Civil Rights Act.

NASA was, however, still far from fully embracing diversity in its workforce. Around the time of the EEO amendment's passage, NASA and fifteen other federal agencies were sued for discriminatory hiring practices in Alabama. Indeed, NASA's EEO office released a report in 1973 calling the agency's affirmative action efforts a near-total failure. Whereas people of color on average constituted 20 percent of the federal workforce, they made up only 5.1 percent of employees at NASA, the lowest representation of all agencies. Women were hired primarily for clerical jobs.[16] The report had been spearheaded by the office's deputy, a Black woman named Ruth Bates Harris, whom administrator James Fletcher ultimately fired for being a "seriously disruptive force." Harris's dismissal did not help NASA's case when agency officials attempted to justify their poor diversity hiring record during a January 1974 Senate hearing. Legislators concluded based on NASA's actions that the agency had an undeniable record of discrimination.[17]

NASA was not quite ready to hire new astronauts in the early 1970s: the shuttle was several years away from flight readiness and its requirements were not fully understood. But NASA's top officials recognized that the social milieu required the agency to select astronauts who would better reflect the nation's demographics while maintaining technical qualifications. Fletcher redoubled efforts to recruit women and diverse candidates and provide them with career growth and leadership opportunities, claiming publicly that the agency was developing plans to get women and members of minority groups into space using the shuttle.[18] Dudley McConnell, NASA's assistant administrator for EEO and highest-ranking Black official, responded to public letters about the astronaut corps's diversity. "We fully expect that flight crews aboard our Shuttle will be representative of all people both male and female—minority and nonminority," he replied to a Virginia man, adding: "We feel that the provision of true equal opportunity is not only our moral responsibility but also our duty; it is something that is right."[19]

The agency explained what it was doing to change its astronaut corps's demographics. In a letter to Gemma Arnott of South Carolina, McConnell said that "the Space Shuttle is specifically being designed to accommodate both male and female flight-crew members." McConnell was undoubtedly referring to the shuttle toilet, which NASA was working to make accessible to members of both sexes. Meanwhile, beginning in 1973, NASA's Ames Research Center recruited women for horizontal bed rest and centrifuge studies to understand the physiological impacts of the gravity changes one would experience during various phases of a shuttle flight. Following similar research conducted with men, the first study involved twelve air force nurses twenty to thirty-five years of age, while a 1977 investigation involved women thirty-five to forty-five years old from a variety of professional backgrounds. The researchers found no inherent issues prohibiting women from flying aboard the shuttle.[20] In both words and actions, NASA was moving toward accepting a broader range of Americans as participants in its shuttle program.[21]

Once committed to flying women and minorities on the shuttle, the agency had to figure out how to take affirmative action in the astronaut corps. NASA public affairs director John Donnelly advised Fletcher in 1974 that "it may be time for us to go out and get ourselves a black astronaut from the Air Force or Navy" as a "visible means of demonstrating NASA's commitment to EEO" and to generate a "favorable response in the minority community and among some of our congressional critics." Donnelly noted that NASA would need to make sure the selected individual was "as well qualified as possible" to "minimize the potential backlash based on the idea that this is a publicity gimmick." He suggested asking military agencies to lend NASA several pilots, including at least one Black pilot, to assist with shuttle flight planning. This way, NASA could evaluate the pilot's qualifications before permanently naming him (Donnelly stated that it would be a "him") to the astronaut corps.[22]

NASA did not follow through with Donnelly's suggestion. But later in 1974, Fletcher requested an astronaut recruitment plan that would consider how to attract Black and female astronauts. Associate administrator for space flight John Yardley worked with Kraft on an approach. NASA's Shuttle Astronaut Recruitment Program kicked off in July 1976 with a press announcement encouraging minority and women candidates to apply. NASA planned to tap Johnson Space Center EEO office head Joseph Atkinson, who was Black, and Carolyn Huntoon, Johnson's highest-ranking female manager, to be part of the selection panels for the new pilot and mission specialist astronauts.[23]

The key to NASA's strategy was to announce its requirement for new astronauts broadly, targeting minority groups and women with the credentials the agency sought in the places it thought it could best reach them. Shortly after NASA released its call for astronaut applications, Fletcher contacted the Tuskegee Airmen to help locate qualified candidates. The agency also interviewed its co-op students and sent letters and brochures to nearly a thousand colleges and universities asking them to recommend potentially qualified female students and students of color.[24]

Several months into the yearlong recruitment period, Alfred Clinkscales, manager of NASA's equal opportunity professional recruiting office, felt that interest by women and members of minority groups had been unsatisfactory. Clinkscales asked recruitment coordinators at each of NASA's field centers to make attracting people of color and women to astronaut positions their highest priority. He sought their ideas for an "all out recruitment effort" and convened a meeting at NASA Headquarters in January 1977 to discuss ways to redouble publicity efforts. The agency's personnel office set an ambitious schedule of visits to colleges and universities with high female and minority group enrollments and continued to seek school officials' referrals of promising students, whom NASA recruiters contacted directly.[25] The agency also contacted ninety aerospace companies to identify qualified candidates. Hispanic outreach consisted of making contacts through the League of United Latin American Citizens and other major Hispanic organizations and magazines such as *La Luz*, appearing at Hispanic conventions, and conducting recruitment campaigns in southwestern states, California, and Puerto

Rico. The agency aimed to reach Black candidates through the NAACP, placing ads in *Ebony* and *Black Enterprise*, and searching names listed in *Black Engineers in the US* and *Who's Who among Black Americans*. Women were sought through the Society of Women Engineers and the American Association of University Women, among other organizations.[26]

NASA found one of its most potent recruitment assets in the realm of science fiction. Nichelle Nichols, who had played Lieutenant Uhura, the communications officer aboard the Starship *Enterprise* on the *Star Trek* television series, proved passionate about the real-life NASA space program and the educational advancements of women and people of color in science and engineering. A Black actress with mixed racial heritage, Nichols had given a speech at the annual meeting of the National Space Institute, where she was a board member. She called attention to the dearth of women and lack of diversity in NASA's astronaut corps, which she attributed to fear, apathy, and a lack of information among these groups regarding opportunities to participate. Recognizing Nichols's credibility and visibility, the agency retained her as a spokesperson in February 1977, hoping she could convince women and people of color that NASA was serious about wanting their participation in the astronaut corps. Over the course of her six-month contract with NASA, Nichols appeared on nearly three dozen major television and radio talk shows geared toward general audiences as well as women and minorities.[27] She also recorded several public service radio spots at the agency's request.

Nichols simply and clearly vocalized NASA's imaginary for the shuttle, asserting that space is for everyone. "There were no social barriers on the Starship *Enterprise* . . . men *and* women . . . all races and ideologies," one public service announcement script read. "That's the way NASA wants it with its new Space Shuttle program."[28] Nichols visited colleges and aerospace companies, seeking to persuade female and nonwhite scientists and engineers that NASA had changed its ways. She explained that NASA was guilty of "ingrained, institutionalized racism" in its previous astronaut selection processes, which were biased toward finding science and engineering expertise that only white males could once attain. But the agency was now offering a "tremendous opportunity" for women and men of color.[29]

FIGURE 6.2. NASA's 1978 astronaut class included the first six women that would fly on the shuttle. From left to right: Rhea Seddon, Anna Lee Fisher, Judith Resnik, Shannon Lucid, Sally Ride, and Kathryn Sullivan. Photograph from NASA.

NASA's recruitment campaign had its intended effect. A total of 8,079 people applied for pilot and mission specialist positions, including 1,544 women and several hundred people from racial and ethnic minorities. More than 20 percent of those who met the mission specialist qualifications were women and 6 percent were people of color.[30] In January 1978, NASA announced that its thirty-five new astronaut selectees included six women and four people of color, demonstrating to the American people that NASA was committed to welcoming highly qualified, socially diverse astronauts. Female recruits Anna Lee Fisher, Rhea Seddon, and Shannon Lucid came from the fields of medicine and the biological sciences, Sally Ride was a physicist, Judith Resnik had worked as an engineer, and Kathryn Sullivan had a background in geology (see fig. 6.2). Guion Bluford, a Black American, was an air force major and an aerospace engineer, Ronald McNair had a physics background, and Frederick Gregory—the

FIGURE 6.3. From left to right: Ronald McNair, Guion Bluford, and Frederick Gregory, NASA's first Black astronauts to fly into space. Photograph from NASA.

only pilot among the women and candidates of color—was also an air force major (see fig. 6.3). Ellison Onizuka, a Hawaiian of Japanese descent, served as a manager and engineer at the Air Force Test Pilot School in Southern California (see fig. 6.4).

The news media lauded NASA's selection of racially diverse and female astronaut candidates as a "humanizing of the astronauts" and approached it with enthusiastic curiosity. The women in particular were peppered with questions. Inquiries of a personal nature surprised and downright irritated some, such as how their new jobs would affect their family lives and how often they played racquetball. But questions about the historical significance of being selected elicited insightful responses. Ride told the *Houston Post*: "I think I owe a lot to the women's movement. I think I came

FIGURE 6.4. Ellison Onizuka, NASA's first Asian American astronaut. Photograph from NASA.

along . . . at an excellent time because the women's movement had already paved the way." Ride's flight aboard STS-7 in June 1983 made her the first American woman in space and represented another leap in the women's movement. Similarly, when Bluford became the first Black American to fly in space, the Associated Press hailed the event as an "overdue break-through in the struggle for racial equality."[31]

During the next selection cycle, which culminated in July 1980, the agency chose nineteen new astronaut candidates from nearly three thousand applicants. Those selected included one Black pilot, two female mission specialists, and the first Hispanic American mission specialist, Franklin Chang-Diaz. And just as the commitment to social diversity continued in ensuing astronaut selection cycles, women and people of color entered significant positions related to the shuttle program on the ground, too. By 1987 women held 10.2 percent of NASA civil service science and engineering jobs. Mission control welcomed its first female shuttle flight director, Michele Brekke, in 1985, and its first Black flight director (for the International Space Station), Kwatsi Alibaruho, in 2005. Trudy Tiedemann became NASA's first female commentator for human space missions, serving as the voice for the shuttle approach and landing tests in 1977, while Kennedy Space Center's Lisa Malone was named NASA's first female launch commentator in 1989.[32]

As NASA continued to welcome new crops of flyers, criticisms arose that the agency did not open the astronaut selection process widely enough. Some members of Congress, the media, and citizens noted that NASA's astronaut selections favored applicants from military branches or its own workforce. While thirteen of the twenty mission specialists named in NASA's 1978 astronaut class hailed from academia, industry, or medical institutions, the classes of 1984 and 1985 were entirely composed of individuals with NASA or military backgrounds. Physician Mae Jemison was the only candidate with alternative credentials in the 1987 class. A letter to the editor of *Aviation Week & Space Technology* expressed the opinion that this phenomenon "should be viewed with alarm by everyone who is interested in America's space program being an open one" and that NASA should reserve equal numbers of slots for outsiders and those connected to NASA and the military. Congressman Manuel Lujan Jr. of New Mexico introduced an amendment to NASA's 1987 authorization for the agency to make a greater effort to select candidates from industry, academia, and other government agencies, but the legislation died for unrelated reasons. NASA nonetheless defended its selections, which continued to be heavily composed of candidates from either NASA or military backgrounds throughout the shuttle's lifetime, on the basis that people who aspired to be astronauts tended to pursue careers in these institutions to gain the sort of experience the agency valued.[33]

Even as NASA officials continued to affirm their commitment to diversity hiring, they tended to meet outside pressure to expand the agency's breadth of astronaut selections with conservatism. The agency rejected an effort by the American Society of Aerospace Pilots, a group formed by United Airlines, to develop a two-year training course to prepare commercial pilots to fly the routine shuttle missions that NASA had asserted would eventually become a reality. According to Johnson Space Center public affairs official Hal Stall, NASA had no plans to hire externally trained pilots because the shuttle "may look like an airplane, but flying it is nowhere as simple."[34] NASA carefully balanced its promise that the shuttle would be more inclusive with its sense of how to remain responsible in managing the program, taking the steps it deemed prudent to maintain political and public support for the vehicle.

Accommodating Outsiders

The agency set another precedent in expanding who could fly on the shuttle by opening seats to people from outside the ranks of career astronauts. In 1973 Fletcher told the *Saturday Evening Post*, "The Shuttle will open the laboratory of the space environment not just to specially trained astronauts, but to engineers, scientists, technicians and others who will be able to accompany their experiments into orbit." Fletcher and other NASA officials embraced the idea that allowing scientific payload owners to fly with their experiments would further incentivize them to use the shuttle and Spacelab for research purposes, boost the return of these payloads, and validate the shuttle's value to the nation. As it was, members of the space science community had formally expressed interest in operating their own research experiments in space beginning with the Space Science Board's 1962 study. Thus, while NASA officials formulated the mission specialist position to integrate flight and payload operations, they also began discussing the concept of the payload specialist—a person who would not join the agency's corps of career astronauts but would fly on a mission to support a specific satellite or experiment. Hans Mark, who was serving as director of NASA's Ames Research Center in California in the mid-1970s, said of the position: "I believe that the Payload Specialist will ultimately become by far the most important member of the Shuttle team."[35]

Beginning in 1974, the agency embarked on a series of studies and meetings to plan for flights of payload specialists. As the first Spacelab mission experiment selections got under way, senior leaders from across NASA worked to develop a policy outlining the selection and training procedures for payload specialists with experiments sponsored by NASA. Officials agreed on some matters. For example, the scientific community should have a hand in choosing candidates, and payload specialists would require less training on shuttle systems than career astronauts. But Kraft had several objections. Kraft understood the scientific and marketing value of the payload specialist position but, along with Slayton, had a vested interest in preserving flight opportunities for NASA's career astronauts. Why, Kraft asked in letters to NASA Headquarters officials, should NASA jump to involve external space flyers when the agency had a cadre of paid-for and trained mission specialists who could be taught to operate researchers' payloads? Kraft believed that NASA should invite payload specialists only to handle highly specialized experiments. He also regarded payload specialists as passengers rather than crewmembers and expressed safety concerns, stating that "a passenger onboard the shuttle poses a special kind of problem."[36]

In March 1978 the agency published in the *Federal Register* its policy and procedures for payload specialists associated with payloads funded by NASA or in partnership with another entity (as Spacelab was). The rule was free of the limitations Kraft advocated. It did not levy conditions for enlisting a payload specialist, stating only that the head of the NASA organization sponsoring the payload would determine the need. If he or she decided that a payload specialist was warranted, then the project team would convene an Investigator Working Group to identify and recommend suitable candidates for the sponsor's approval. The selected candidate would need to meet NASA's class III space flight medical selection standards—a less rigorous set of qualifications than was required of mission specialists—and undergo training to operate the payload and become familiar with shuttle systems.[37]

Two months later, Kraft wrote to NASA deputy administrator Alan Lovelace expressing frustrations about the selections of payload specialists then in progress for the first and second Spacelab missions. He lamented that the finalists identified by Investigator Working Groups had not risen to the top when the same individuals applied for mission specialist positions in NASA's 1977 call. Moreover, only two of the experiments required specialized expertise, and he claimed that the nominators did not adequately consider whether mission specialists could serve their needs. While acknowledging that the scientific community's involvement in choosing payload specialists added credibility to NASA's commitment to researchers' participation with the shuttle, Kraft opined that "we are becoming too enamored with the public relations aspects of such selections without properly considering what we are doing practically and operationally." Operations and program marketing, according to Kraft, "are not necessarily compatible factors and must be carefully integrated to insure mission success." He recommended that NASA pull together its operational, scientific, and marketing groups to determine how to safely fly the shuttle while accomplishing scientific objectives at the lowest possible cost to the user.[38]

Later that year, NASA nonetheless announced Byron Lichtenberg, an engineer at MIT, and University of California physicist Michael Lampton as the first American payload specialist candidates. The European Space Agency chose Swiss astronomer Claude Nicollier, Dutch physicist Wubbo Ockels, and German physicist Ulf Merbold to represent Europe. Lichtenberg and Merbold became the first non-NASA astronauts to fly aboard the first Spacelab mission, on STS-9, in November 1983. But in response to Kraft's concerns, NASA administrator Robert Frosch ordered a review of the policy, which was amended in 1980. The revised rule stated that NASA would provide payload specialists opportunities to support mission payloads when a specific requirement warranted their expertise. Otherwise, payload specialists would be selected from NASA's mission specialist cadre.[39]

The policy change did not deter some NASA officials from continuing to capitalize on the payload specialist position to make space flight opportunities available to a broader cross-section of people to boost the shuttle's vitality and viability. Yardley, for one, believed that offering corporations the opportunity to choose their own representatives to fly with their payloads would incentivize them to use the shuttle. While Frosch was evaluating the guidelines for NASA-related payloads, NASA's Office of Space Flight considered the possibility of allowing external customers to select their own payload specialists.[40] Although Frosch had sided with Kraft on limiting the use of payload specialists, his successor during the Reagan administration, James Beggs, shared Yardley's position. Beggs liked the idea that NASA could give companies flying satellites aboard the shuttle this incentive. In October 1982, Beggs informed NASA's House and Senate appropriations and authorization committees that the agency's policy on payload specialists was "overly restrictive" and that he intended to expand it to domestic and foreign shuttle customers. Beggs added that he planned to extend opportunities for scientists to fly and for Department of Defense payload specialists to accompany security payloads.[41]

Shuttle customers soon put NASA's willingness to include payload specialists to the test. McDonnell Douglas executive Jim Rose approached NASA shuttle program manager Glynn Lunney in 1982 about flying a company representative with its electrophoresis experiment. According to Charles Walker, the McDonnell Douglas engineer who would ultimately fly three times with the equipment, Rose told Lunney: "We really would learn the most we possibly can and more than we can do with a mission specialist if we get the opportunity to have a payload specialist devoted specifically to the electrophoresis device and its research and development activities during a flight." Lunney encouraged Rose to put in a formal request, and in early 1983 NASA approved Walker as the first corporate payload specialist (see fig. 6.5).[42]

Beggs wanted to revise the rule to allow payload specialists to fly whenever a customer deemed it necessary, regardless of what NASA thought of a payload's uniqueness. Although the agency never changed the federal regulation, it endorsed the requests of major customers such as McDonnell Douglas that asked to fly a representative with their payloads. Customers included companies and foreign governments that reimbursed the agency to deploy communications satellites from the shuttle's cargo

FIGURE 6.5. Payload specialist Charlie Walker operates McDonnell Douglas's Continuous Flow Electrophoresis System on STS-41D aboard space shuttle *Discovery*. Photograph from NASA.

bay and wanted to conduct research that interested them. They selected their own payload specialists and paid NASA the marginal cost for their flights and training (Walker's first trip cost McDonnell Douglas $40,000). RCA's Robert Cenker, for example, flew as a payload specialist aboard STS-61C in 1986 to observe the deployment of the company's Satcom K1 satellite and to test a classified infrared camera for an air force client. Saudi Arabia sent Sultan bin Salman Al-Saud to accompany its Arabsat communication satellite on STS-51G in June 1985, photograph the Arabian Peninsula from orbit, and conduct research on fluid behavior in microgravity that would potentially be useful to the oil industry.[43] Following Walker's first flight in 1984, nearly every shuttle mission up through the catastrophic launch of *Challenger* in January 1986, and many subsequent missions, included one or more payload specialists from industry, academia, or the military.

While the payloads and experiments scheduled for each flight determined the need for most payload specialists, diplomacy and politics played a role, too. President Ronald Reagan welcomed nations including France, Italy, China, and Brazil to name delegates to fly aboard the shuttle apart from any previous dealings with NASA or plans they had to send payloads into space. It was a gesture of international goodwill as well as an effort to incentivize foreign nations to put payloads on the vehicle. The move paralleled the Soviet space program's practice, in place since 1978, of selecting "guest cosmonauts," primarily from Communist-bloc nations but also from the West, to visit the first Soviet space station, *Salyut*.[44] French fighter pilot Patrick Baudry flew aboard the shuttle in 1984 and others would follow in response to Reagan's invitation.

Beggs similarly recognized that granting key members of the US Congress access to the shuttle could clinch their support for human space flight endeavors. He invited the chairs of NASA's appropriations and authorization committees in the House and Senate to fly aboard shuttle missions as part of their oversight duties. The action was prompted when in 1981 Edwin Jacob "Jake" Garn, a former navy aviator who served as the chairman of the Senate subcommittee that doled out NASA funding, petitioned Beggs to fly aboard the shuttle. The NASA administrator knew that the agency needed to get the shuttle to operational status first, but he took Garn's request very seriously. "'If he wants to fly, by God, he'll fly,'" Beggs recalled saying to himself. "I thought if he flew then he could come back and he could really make it real for the other members of Congress.[45]

In naming Garn as a payload specialist in early 1985, NASA stated that the experience would allow the senator to share the knowledge he gained and aid in the decision-making of those in Congress and "in turn, benefit the American taxpayer." The agency further justified the flight by pledging that Garn would "perform some meaningful task."[46] They assigned him to serve as a subject for several medical experiments on STS-51D, which flew in April 1985. Florida's Bill Nelson, chairman of the House subcommittee overseeing space issues, also accepted NASA's invitation and flew on STS-61C in January 1986, the mission preceding the final flight of *Challenger*. Not all Congressional leaders accepted the offer, however. Don Fuqua, chairman of the House Committee on Science and Technology, rejected the opportunity on the premise that NASA should fly only trained astronauts and that he could not focus adequately on both flight training and serving as a member of Congress.[47]

Nelson's constituents living near Kennedy Space Center expressed strong support for his participation in a shuttle mission. Many journalists, however, saw flights by politicians as a misuse of the shuttle. Florida media critics complained that these arrangements constituted "the ultimate junket" at taxpayer expense and meant that a member of Congress "no longer had the arm's length relation with NASA that is needed to ensure that he oversees the space program objectively." Others noted that these flights would restrict "legitimate research" and be unfair to astronauts who had trained for years.[48] Clearly, what constituted credible uses of the shuttle was debated, and NASA officials constantly considered whose participation would help to ensure the vehicle's viability.

Indeed, the presence of members of Congress and other payload specialists on flights was an adjustment for NASA's career astronauts. According to shuttle pilot Hank Hartsfield, "Some of the guys who hadn't flown resent these guys coming along and these people flying. 'I'm sitting here, I'm trained, I've been picked, and these people are going.' So it's a morale issue for them." Some astronauts worried that a payload specialist's limited training—Garn's experience of less than three months being an extreme example—meant that a crew could never be sure of the individual's level of preparedness for flight. "You had to determine their personality. It was hard to do in a short period of time. Were they going to be stable?" Hartsfield pointed out. "If you had a big problem, you could wind up having a person that wasn't used to that kind of conditions to endanger the rest of the crew because you have to attend to them."[49]

Walker, who described himself in his payload specialist role as "the itinerant space flyer," explained the skepticism he endured. A few people inside and out of the Astronaut Office, he said, told him: "'You're not one of us. You're along for the ride, and you've got a job to do.'" While Walker never experienced overt belligerence from career astronauts, he noted that there was "no clearer indication" that payload specialists were "outsiders" than the confinement of their work spaces at Johnson Space Center to separate buildings. "It was made clear to us from the beginning that they didn't expect to see us over [in the Astronaut Office] in Building 4 except for scheduled meetings," Walker said. Even with the need to ensure that the payload specialists would work effectively and safely as part of mission crews, there was still a tension in the Astronaut Office and among mission managers about not wanting "to bring them into the office too close."[50]

NASA's highest level of leadership nonetheless remained committed to making payload specialists a regular presence on flights, and over time career astronauts grew more comfortable with their inclusion on shuttle missions. Having disparate backgrounds and objectives, trained payload specialists proved able to function aboard flights without putting crewmembers in danger, and their presence freed the career flyers from having to learn to operate customers' payloads. By the time of the *Challenger* accident, the payload specialist office was unified with the Astronaut Office, and, according to Walker, only "a vanishingly small number of people" held separatist attitudes. "[It] took a couple of years," Walker reflected, "and the adaptation began to happen, I think, in serious fashion as [they] saw . . . you're taking the risks and you're flying in this thing together. It's not a positive to keep yourselves apart during the preparations for all of that. So I think that reality certainly came home."[51] The shuttle had managed to bring new flyers into the fold, building much goodwill along the way.

Welcoming "Ordinary" Citizens to Space

From its earliest days, NASA received letters from American citizens volunteering for astronaut duty or asking when the agency would take "ordinary" people on trips to space. Some ninety-three thousand people had added their names to a waiting list started by Pan Am Airways, which in 1964 announced that it would one day offer passenger flights to the moon. The notion of routine space flight had also been featured in science fiction literature for decades and was reignited in television series such as *Star Trek* and *Battlestar Galactica* thanks to Apollo's influence. As shuttle development got under way and NASA officials spread the idea that the vehicle's frequent flights and roomy interior would make space more widely accessible, letters continued to roll in. When Hugh Downs of ABC News asked Low about his chances of flight as a news correspondent, Low replied: "I sincerely hope that the day when it will be possible for a journalist to go along on a Shuttle mission as an observer is not too far off."[52]

While NASA officials knew that the shuttle would not be able to serve every person who dreamt of going into space, some agency personnel believed that the new vehicle would make strides in realizing this vision. Moreover, broadening participation beyond scientists, engineers, and military pilots could have an effect even on those who lacked space flight aspirations in that it could make the enterprise more relatable. The Mercury, Gemini, and Apollo astronauts had ably pioneered the space frontier but had been less proficient in describing the experience to the billions below on Earth. Apollo 16 astronaut John Young admitted after his mission that having an artist along would have helped. "There were things I saw and felt," Young said, "that just can't be put into words." An *Aviation Week & Space Technology* reader agreed, sending a letter to the editor noting that "as long as space is perceived as a sterile laboratory where high-tech eggheads measure stuff, the potential public excitement over NASA's goals will remain squelched."[53]

In the mid-1970s, multiple sources of external pressure spurred NASA officials to look seriously into flying people who would help to humanize the human space flight program. After receiving a briefing from NASA on the agency's shuttle flight test plans, the National Research Council's Aeronautics and Space Engineering Board stated that the proposed flights "were not exciting, lack-lusters, and had nothing to stir the imagination of the American public." Philip Culbertson, manager of NASA's Advanced Manned Missions Planning Group in the Office of Manned Space Flight, held a brainstorming session in February 1976 to come up with ideas that might capture public interest and attention. When Culbertson shared his list with Low, the deputy administrator endorsed Culbertson's suggestion of flying a "unique personality" outside NASA's astronaut ranks on the shuttle.[54] The following month, a *National Geographic* editor contacted Fletcher about flying a journalist on the shuttle to help NASA communicate the space flight experience to the public. A decade earlier, the Martin Company had recommended to NASA to send a *National Geographic* photojournalist on a Gemini mission to boost public interest. The idea resonated with Fletcher, who replied that NASA would give consideration to the editor's interest when the opportunity arose. As a first step toward that future, Fletcher forwarded the editor's letter to public affairs official Bob Shafer with a suggestion to start an "observer" program for shuttle flights.[55]

A group led by shuttle program official John Hammersmith and including representatives from NASA's human space flight, public affairs, legislative, and legal offices set out to study the possibility. Their July 1976 report evaluated people with various types of careers and experience—*National Geographic* photojournalists, news media representatives, students, entertainers, eminent scientists, and "the layman." Collectively, they favored a journalist to provide broad coverage of the shuttle mission consistent with the Space Act's mandate to NASA to disseminate information widely about the space program and also to establish that space flight was no longer the province of those from elite professions. They also suggested making flight opportunities available to people from other walks of life who would be recommended by committees of their peers. In 1978 NASA's public affairs office drafted a policy for shuttle passenger selection, justifying it on the idea that passengers could "contribute to a greater understanding of space flight" by sharing their experiences broadly. Because the plan emerged, however, when the payload specialist policy was under internal review and the shuttle faced development delays, the agency shelved it for the time.[56]

Policy or no policy, as the shuttle's launch debut approached, NASA officials together with the media continued to promote the idea that the shuttle would provide access to space for almost anyone. In December 1979, *Parade* magazine featured a story on the shuttle in which former external affairs chief Herb Rowe was quoted as saying that the agency had a list of celebrities, scientists, journalists, and politicians it would invite to fly, and shuttle spokesperson David Garrett said that the first nontechnical participant "is sure to be a journalist." Tom DeVries told readers that those who would fly as payload specialists would "have to meet qualifications that are very low, quite within the reach of armchair command-module pilots like us." A February 1977 article in *American Legion Magazine* suggested that anyone could be a shuttle passenger, noting that "outings in orbit may be practical a lot sooner than you think." Due to this hype, requests from citizens across the country to fly aboard the shuttle began to flow in to NASA by the hundreds. Public affairs director Robert Newman believed the news media was exerting "too much external pressure" for NASA not to develop a passenger policy soon.[57]

Less than a year after the shuttle's first orbital test flight, Beggs was ready to respond. Like Fletcher before him, Beggs believed that citizens rightfully had a future in space that the shuttle could at least partly accommodate. His sense that NASA ought to consider flying people who were not trained as astronauts was affirmed after he found that other exploration initiatives had involved nonprofessionals in their ranks. Naval officer and explorer Richard Byrd, for one, had taken a Boy Scout on his first expedition to Antarctica. Beggs wrote in a NASA employee newsletter that "a new opportunity is emerging—for people to go themselves, to see for themselves and to share with others." In February 1982 he established a task force of the NASA Advisory Council to study options for selecting private citizens for shuttle flights. He told the task force at its first meeting that "NASA's objective is to maintain the openness of the program and to invite the public to participate to the extent possible." Beggs said that he did not want NASA to focus solely on selecting national celebrities for flight because then the agency "would be charged with simply promoting the program which is not our intent."[58]

The task force was chaired by former NASA chief scientist Naugle and included former astronaut Richard Truly, former NASA public affairs chief Julian Scheer, novelist James Michener, historian Sylvia Fries, marketing executive Florence Skelly, and aerospace executives Daniel Fink and Willis Hawkins. For nearly a year, the task force studied and debated the merits of a citizens-in-space program, who should be tapped to fly, and how NASA should go about selecting them. Skelly advocated that NASA fly "key special people," including communicators well versed in describing scientific and technological issues. Such individuals, Skelly argued, would be valued by policy leaders and citizens interested in NASA, whereas flying "the man on the street" could invite external criticism as a waste of taxpayer funds. Fries, meanwhile, worried less about critics and preferred a more democratic route: a lottery open to all interested parties.[59]

In June 1983 the task force presented NASA with its conclusion that the "flight of private citizens is both feasible and desirable." Making the American citizenry an integral element of the program, they said, would accomplish three things: it would expand human knowledge, contribute to human culture, and educate the broader public about space program activities. The task force thought that the agency could justify such an initiative as long as citizen flights fulfilled purposes outlined in the Space Act—namely, to preserve US space leadership and to disseminate information widely. Inviting citizens as tourists would not be justifiable. The group refrained from recommending a specific approach to selection but suggested that after flying a few people from categories that would serve the Space Act's aims, NASA could consider "opening space flight to all people."[60]

NASA's leadership endorsed the task force's recommendations and moved toward figuring out how to implement them. The agency established a working group including staff from the public affairs, legal, space flight, science, and equal opportunity offices. For years, NASA had rejected requests from citizens to fly aboard the shuttle. Now the agency had adopted a requirement to increase inclusiveness in human space flight to connect the public more closely with the program. The working group needed to develop equitable selection procedures without compromising crew safety or marring the agency's credibility. It was not an easy task. NASA spokesman Brian Welch summarized the challenge: "How do we pick individuals without making other people mad? How do we make sure we haven't cheapened the space program?"[61]

The draft policy that NASA published in the *Federal Register* in December 1983 aimed to balance openness and fairness with seriousness of purpose and safety. It described the rule's intent as "to increase the access of the general public" to the shuttle and communicated NASA's commitment to "the participation of a wide and diverse array of participants, including women and minorities." It noted that NASA's desire was to provide citizen space flight opportunities as long as the flights met purposes defined in the Space Act and posed no threat to mission safety or success. In other words, flying people who could help promote understanding of space flight would be acceptable, but allowing Michael Jackson to perform his moonwalk in the shuttle payload bay—as had been proposed by someone who called NASA claiming to be his agent—would not be. Even as NASA officials promoted the shuttle's commercial value and candidacy for eventual privatization, they were determined to avoid accusations of conducting frivolous publicity stunts, just as they did with the payloads the shuttle would carry.[62]

Other aspects of the proposed rule demonstrated NASA's intent to maintain careful control of who flew. Participants would need to be free of medical conditions that could impair their performance on a mission. They would have to undergo background investigations and training and would have to sign agreements with NASA concerning insurance, liability, and rights to outside compensation. The agency would announce opportunities for people from specific professions it would consider for flight. NASA would designate outside "peer group" panels to screen applications but would retain final selection authority. Several of the twenty-two people who commented on the rule voiced their opposition to this approach. They, and others who wrote to NASA on separate occasions, argued that NASA would most effectively open space to the public by flying "common citizens" selected via random lotteries.[63] The input did not sway the agency.

One additional point raised by the policy regarded what to call these new kinds of space flyers. The draft referred to them as "citizen observer/participants." A public commenter, however, suggested including "astronaut" in the title chosen. But career astronauts were sensitive to the introduction of citizen shuttle passengers, just as they first looked unfavorably at payload specialists. Some lamented that these new flyers could delay their own assignment to missions or risk crew safety. Mission specialist Mike Mullane, for one, said the notion that ordinary citizens could become shuttle astronauts was "immoral" when the shuttle was still largely an experimental vehicle. The agency decided to reserve the title "astronaut" for professional crew members and settled on the name "space flight participant."[64]

NASA finalized the policy in April 1984, nearly verbatim with the proposed version, marking the start of what would be known as the Space Flight Participant Program. Alan Ladwig, the longtime advocate of public space flight who had been running the Shuttle Student Involvement Program, became its manager. Beggs felt that as long as the shuttle flew fifteen to twenty missions per year, NASA could accommodate two to four citizen flyers annually.[65] But whom would the agency choose? The regulation did not name specific groups. Since starting to publicly discuss the possibilities the shuttle might present for citizens, NASA had received thousands of letters and phone calls from people asking if they could be the first of their kind to fly into space, whether a poet, artist, evangelist, child, Olympic athlete, or disabled person. Some of those who had signed up for a Pan Am flight to the moon asked NASA about redeeming their reservations with a shuttle flight.[66] Inquiries about flying came from celebrities including singer John Denver, the Rolling Stones, and newsman Walter Cronkite.

The year before NASA finalized the Space Flight Participant Program policy, a national commission had issued a report called *A Nation at Risk* that condemned the state of the American education system. The report set the Reagan administration scrambling to identify ways to improve academic performance, and NASA began working with the White House to formulate Operation Liftoff to create space-focused programs and materials to boost science and math skills in K–12 classrooms. When NASA's White House liaisons asked about flying a teacher aboard the shuttle to captivate student interest, NASA's committee for Space Flight Participant Program implementation advised Beggs to select an educator to serve as NASA's first citizen space flyer. Beggs appreciated that flying a teacher would support Reagan's commitment to improving science education while meeting NASA's goal to normalize human space flight and enhance the shuttle's societal relevance. "The biggest receptive audience we have in this country are the kids," Beggs said. "Kids love space. A teacher could give you an introduction to those kids that no one else could." At Beggs's recommendation, President Reagan publicly directed NASA on August 27, 1984, to make a teacher the agency's first competitively selected citizen in space.[67]

That fall, NASA partnered with the Council of Chief State School Officers to develop selection criteria and process applications during the 1984–1985 academic year for what became known as the Teacher in Space Project. Announced in November 1984, the initiative invited elementary and secondary schoolteachers from public and private schools in all US states and territories as well as in Department of Defense overseas schools and Bureau of Indian Affairs schools to submit applications by February 1, 1985.[68] NASA advertised the program extensively, conducting outreach to education-focused organizations and attending relevant conferences. Ladwig even appeared on the *Late Night with David Letterman* show to promote the opportunity.

The applications were not for the faint of heart: at fifteen pages long, they required teachers to submit letters of recommendation, write essays outlining their education philosophies, and explain the educational projects they would conduct aboard a shuttle mission. NASA distributed more than forty-five thousand applications. More than eleven thousand teachers completed applications, which they often sent accompanied by letters, drawings, and paintings from their students who endorsed their selection.[69] Each state, territory, and agency identified two finalists, all of whom were invited to Washington, DC, for a week in June 1985 to receive briefings on NASA programs and participate in interviews by a panel

FIGURE 6.6. Christa McAuliffe (*left*) and Barbara Morgan were selected as Teacher in Space and backup for the position. Photograph from NASA.

that included Apollo astronaut Gene Cernan, rocket scientist Konrad Dannenberg, and Pam Dawber from TV's *Mork & Mindy*.

NASA narrowed the field to ten national finalists. On July 19, Vice President George H. W. Bush announced the selection of New Hampshire social studies teacher Christa McAuliffe as America's first teacher headed to space. Idaho teacher Barbara Morgan would be McAuliffe's backup (see fig. 6.6). Assigned to the STS-51L *Challenger* mission crew, McAuliffe was slated to telecast live classroom lessons to the nation's schoolchildren. The first lesson, "The Ultimate Field Trip," would orient students to the shuttle, and "Where We've Been, Where We're Going, Why?" was to explain the benefits of human space flight. Classrooms with access to satellite dishes or cable networks carrying the NASA Select television channel could also watch coverage moderated by Morgan throughout the mission.[70]

One unanticipated benefit of the Teacher in Space Project cen-tered not on the teacher who would fly but on those who would not. At NASA's request, the other eight national finalists took yearlong absences from teaching to support the agency's education initiatives and to assist McAuliffe and Morgan with the development of lesson plans and activ-ities to be carried out on the shuttle flight. Moreover, Teacher in Space Project officials designated all of the state finalists "NASA ambassadors" and encouraged them to conduct public outreach activities in their states and communities. Patricia Palazzolo, who was one of Pennsylvania's two finalists, recalled NASA officials appealing to her and the other teachers to figure out how to make space flight interesting for students in the classroom and for the general public. "It was communicated to us from the very beginning," Palazzolo explained, "as an opportunity to be the ultimate resource of everything people could want to know" about NASA and space exploration. The state finalists reached more than four million people by participating in more than 2,200 public and school-focused lectures, workshops, and other events by July 1986, strengthening con-nections among the shuttle, the education community, and the broader American public.[71]

As the Teacher in Space Project progressed, Beggs became increasing-ly comfortable with accepting citizens as shuttle passengers. He moved forward with plans for a second space flight participant, announcing in October 1985 that journalists would finally get their chance to be consid-ered for a shuttle mission. But what sort of journalist would make the cut? A newspaper editor? A major television network broadcaster? A science writer? A freelancer? NASA teamed with the Association of Schools of Journalism and Mass Communications (ASJMC), made up of 170 journal-ism schools and having connections to dozens more journalism-related professional organizations, to set the selection criteria and evaluate the applications. Emulating the procedure used for the Teacher in Space Proj-ect, NASA and ASJMC developed a comprehensive application requiring recommendation letters and essays explaining how the journalists who aspired to fly believed their participation would serve their profession and the public.[72] By the January 15, 1986, deadline, ASJMC had received more than 1,700 applications. ASJMC's member schools were to choose forty semifinalists in each of five geographic regions, after which a national panel of retired journalists, academics, and others would narrow the field to one finalist from each region. Then NASA would evaluate and select a winner and a backup. NASA expected to fly the winning journalist in the fall of 1986.

Beggs, Ladwig, and others believed firmly that after a journalist, NASA's Space Flight Participant Program would invite many more types of people to take the journey of a lifetime aboard the shuttle and change the way Americans related to NASA's human space flight enterprise. Ladwig was thinking that an artist might be next and had begun conversations with staff at the National Endowment for the Arts about the selection process. The agency was also gearing up to study the feasibility of flying people with disabilities. Bob Walker, a member of the House of Representatives Committee on Science and Technology, believed that flying a person with a physical disability could "prove that those who bear the lifelong burden of a handicap on Earth may be freed to become highly valuable and fully productive" in space. NASA's 1986 authorization bill directed NASA to initiate a feasibility study "to ensure flight opportunities for a diverse segment of the American public, including a physically disabled American."[73] Many people took note of the congressional interest and contacted NASA about such opportunities. Ladwig and legislative affairs assistant administrator John Murphy responded to say that while the shuttle was not then equipped to accommodate people with disabilities, the agency was hopeful and confident that they would one day have the chance to apply for a shuttle flight.[74]

The possibilities for expanding citizen space flight seemed endless. A nonprofit organization founded to promote educational excellence and various uses of space with the backing of Chevron, Lockheed, and other companies proposed a program to fly students along with teachers aboard the shuttle to conduct experiments. And while NASA had no intent to operate a tourism service, the shuttle had potential as a starting point. If the shuttle became a commercial venture—a possibility NASA and the Reagan administration were contemplating—the operator could sell tickets and conduct a lottery to select passengers. By 1985, private companies had proposed developing modules that could accommodate dozens of extra seats in the shuttle's cargo bay. Rockwell put forward a concept to carry astronaut crews to construct a space station one day, and exotic vacations purveyor Society Expeditions hoped to team with NASA to fly tourists using the shuttle until privately owned space vehicles were available.[75] As far as NASA officials were concerned, the shuttle imaginary of accessibility was becoming reality, and it was just a matter of time to see what opportunities emerged. But no one could foresee the disaster that was about to upend NASA's aspirations.

7

Reevaluating the Democratic Imaginary

With the shuttle program's debut, NASA revolutionized its stance on public engagement with human space flight, offering a wide range of people many opportunities to connect with the vehicle, often in very direct ways. Bringing to life the imaginary of a broadly accessible vehicle had not been without resistance from some within the agency and elsewhere who believed that focusing heavily on public involvement would diminish access to the shuttle by those already tied closely to the space program or would somehow suggest that NASA lacked a seriousness of purpose in its endeavors. But NASA officials and advocates who believed in the vision and its necessity—whether to gain public support for or to serve the public with human space flight, or to fulfill some combination of the two—made significant strides toward realizing it.

After NASA had operated the shuttle for a few years, proponents across the agency had many reasons to believe that their mix of strategies for engaging the public with the vehicle was proving effective in galvanizing interest and helping to erase earlier doubts about the human space program's value. In the first five years of shuttle missions, 60 to 70 percent of Americans approved of the shuttle as a worthwhile expenditure.[1] The agency's efforts to share the shuttle resonated with people in many palpable ways. Schoolchildren interacted with the astronauts and participated in initiatives like Young Astronauts and the SEEDS project. Primary and secondary school educators praised the shuttle as an inspirational tool for teaching science, math, and other subjects. People from all walks of life saw potential for the shuttle to carry payloads to space that would help to answer research questions of value to society and of interest to them personally. By the mid-1980s space advocacy groups were proliferating, some spurred on by what Michael A. G. Michaud called the shuttle's "symbolic beginning of the democratization of space."[2] And just as Apollo had hastened a litany of space-related television shows, so too did the shuttle permeate American popular culture, making an appearance in space-themed movies such as *SpaceCamp* and novels such as James Michener's *Space*.

But just as NASA was realizing the indelible mark its efforts to open the shuttle had made on so many Americans, several factors would soon challenge the robustness of the shuttle's democratic imaginary and the agency's willingness to maintain a commitment to broadening opportunities for public involvement. Five years into the shuttle's operational period, NASA and the nation were jolted by the 1986 *Challenger* launch accident. The *Challenger* tragedy revealed that the risks of flying the vehicle were more pronounced, yet were being far less effectively managed, than NASA had indicated publicly and to its political stakeholders. NASA's identity as a bastion of technoscientific expertise and the future of human space flight hung in the balance following the shuttle disaster. Although the participatory imaginary had been instrumental to demonstrating the shuttle's value, NASA backed away somewhat from it as its leaders endeavored to regain the trust of critics, especially in Congress and the media, that the agency remained technically competent and committed to shuttle flight safety. Meanwhile, the accident resulted in a backlog of payloads, leading the agency to reevaluate what sorts of projects and people it would prioritize for flight aboard the shuttle. In addition, a few years after the shuttle had returned to flight NASA received the approval it had long sought from Congress to commence development of an Earth-orbiting space station. With this change, the agency repurposed the shuttle to fulfill its original intent of supporting the station. Would NASA need to justify the vehicle via the democratic imaginary any longer?

The *Challenger* accident and subsequent choices and changes—some within NASA's control and others not—shaped the agency's dedication to its original vision of the shuttle as a broadly accessible vehicle. Officials struggles as they worked to balance serving conflicting perceptions of public expectations, along with those of political stakeholders, the science community, and the commercial space industry, as they aimed to maintain support for pressing forward with the shuttle program and human space flight generally. NASA continued efforts to engage Americans through messaging about the shuttle's value as well as through visual and virtual means. But pressures on the program, including reassignment of the shuttle as a workhorse to service the space station, led NASA to make changes to how far it would go in opening the shuttle to wide use and new flyers. The democratic spirit of the shuttle never fully ebbed, although broadening the direct involvement of a variety of people with the vehicle became less central as NASA's need to prove the space transportation system's viability waned.

Revisiting the Shuttle Imaginary after Challenger

Since the shuttle's development years, NASA had strived to make the vehicle accessible for many sorts of users and uses. The effort, however, had been neither easy nor resoundingly successful. NASA strained to generate enough payload business to realize its economic promises for the shuttle, a situation that was not helped by the Reagan administration's rising concerns about the vehicle's competitive threat to nascent private launch companies. Weather issues and technical challenges with processing the vehicles in rapid succession also had interfered with NASA's ability to achieve anywhere near the flight rates it had touted publicly. Despite efforts to engage the media, many commentators, expecting more from NASA, criticized the shuttle's underperformance. Historian of technology Alex Roland, for one, penned a critique in *Discover* magazine calling the shuttle a "turkey" based on the program's cost.[3]

Turning around these perceptions was critical for NASA. Officials wanted to maintain political support for the shuttle while working to secure funding for a permanent Earth-orbiting space laboratory. President

Ronald Reagan had announced his support for Space Station *Freedom* in 1984. NASA, however, faced difficulty gaining congressional backing for its projected $8 billion price tag, and agency officials intensely felt the need to prove the human space flight program's value. By April 1985, the agency had released a plan to conduct forty-one shuttle missions in thirty-three months—higher than any flight rate it had achieved.[4] Shuttle officials hoped doing so would attract commercial entities that might otherwise entrust their payloads to Europe's Ariane and show naysayers that the agency was indeed turning the corner with the shuttle program and increasing mission availability. Confident in its progress toward making space flight routine, the agency assigned its first "citizen" space traveler, teacher Christa McAuliffe, to the crew of STS 51-L *Challenger*, slated for a January 1986 launch. McAuliffe's presence along with a crew that included Black astronaut Ronald McNair, Japanese American astronaut Ellison Onizuka, Jewish astronaut Judith Resnik, and Hughes Corporation payload specialist Gregory Jarvis suggested that the agency was on its way to achieving broad accessibility for the shuttle.

The very mission that was seen as an epitome of NASA's commitment to the democratization of human space flight would soon challenge the sustainability of that vision. The overnight temperatures preceding *Challenger*'s January 28, 1986, flight were expected to be lower than they had ever been for a shuttle launch. Engineers at Morton Thiokol, the firm that produced the shuttle's solid rocket boosters, expressed concern about the cold weather's effect on the performance of the rubber O-rings that sealed the joints of the rocket segments and kept the fuel and gases from escaping. Shuttle propulsion system managers at Marshall Space Flight Center in Huntsville nevertheless recommended to NASA Headquarters to proceed with the launch, feeling confident that their experience to date with the O-rings made the risk acceptable.[5] Moreover, holding up the launch, which had already endured several delays, would ripple through the aggressive shuttle manifest. It would also likely induce further press commentaries ridiculing NASA, like Dan Rather's opening on *CBS Evening News* on the evening of January 27: "Yet another costly, red-faces-all-around Space Shuttle launch delay due to a bad bolt on a hatch and poor weather."[6] The pressure to launch was intense.

FIGURE 7.1. The breakup of the space shuttle *Challenger* seventy-three seconds after launch on January 28, 1986, resulted in plumes of exhaust from the two solid rocket boosters flying off uncontrollably, contrails from falling pieces of debris, and an expanding ball of gas from the external tank. Photograph from NASA.

The *Challenger* mission resonated with many Americans and created a great deal of excitement. Thousands of spectators, many teachers among them, flocked to the cape on that unusually chilly Florida morning to see the astronauts off on their journey. Countless more, including millions of educators and their classrooms of students, tuned in to view the nationally televised broadcasts of the launch. Seventy-three seconds into *Challenger*'s ascent, they all watched in confusion and horror when the vehicle broke apart and the crew of seven perished (see figs. 7.1 and 7.2).

As investigators would soon discover, NASA's judgment had faltered: the O-rings had not withstood the atypically cold temperatures and failed to seal the boosters to ensure safe flight. The agency had endured

FIGURE 7.2. Spectators at Kennedy Space Center react after witnessing the destruction of the space shuttle *Challenger*. Reproduced by permission from Associated Press.

catastrophe previously when the three Apollo 1 astronauts lost their lives during a spacecraft test on the launchpad in 1967. But NASA's ultimate success with Apollo 11 had had what astronaut Joe Allen calls an "afterburner effect," bestowing on the agency an image around the world as a "can-do agency, the best agency in the federal government."[7] In promoting its imaginary of a highly accessible shuttle, NASA had drifted from publicly discussing the notion that space flight was still a risky enterprise and that calamity could strike again. The agency had been portraying the shuttle as a well-understood technology when in fact it was quite the opposite. Hence, the *Challenger* disaster, which had unfolded on live national television before millions of witnesses, felt all the more tragic because it defied people's sensibilities of what the shuttle, and NASA itself, seemed to promise and embody. It was completely unexpected to those outside of NASA, including the media, elected officials, and many at the agency whose jobs entailed promoting and publicizing the shuttle's success. On that fateful day, one *Chicago Tribune* headline proclaimed, "NASA wizards' legendary infallibility blew up with [the] Shuttle."[8]

Indeed, while NASA had been focused on proving the shuttle to be a viable human space flight initiative and selling the space station to Congress, it suddenly found its reputation of competence and technical expertise imperiled by the *Challenger* disaster. Almost immediately, the agency became the butt of numerous jokes. "What does *NASA* stand for?" went one prevailing wisecrack, which invited the cynical response: "Needs Another Seven Astronauts." Haynes Johnson of the *Washington Post* called NASA "a once-proud agency rudderless, uncertain, torn by dissension and low morale" and noted that "potentially irreparable harm is being done to public confidence" in the agency.[9]

The situation was exacerbated as NASA, shocked by what had occurred, abandoned its longstanding commitment to open communications. NASA's public affairs plan in the case of a shuttle "contingency," or accident, allowed for impounding press cameras and film onsite at Kennedy Space Center to use in investigating the mishap. But the plan stated that the agency would make a statement about the crew's welfare within twenty minutes of the incident. NASA angered and bred distrust among news reporters, many of whom were covering the *Challenger* mission as novices to space affairs, when it impounded their equipment and waited nearly five hours to host a press conference, saying very little when it did. NASA public affairs and shuttle program staff fell silent, failing to return calls and prompting journalists to seek information by other means.[10] Reporters talked to anonymous NASA sources in bars and went to navy ports to see what shuttle remains were salvaged. They demanded NASA's release of information through the government's Freedom of Information Act.[11] Reporters felt misled when NASA issued a statement asserting that the crew had died instantly while evidence ultimately revealed that at least some of the astronauts sensed the danger and survived until the crew module hit the Atlantic Ocean.[12] Michael Cabbage, a space beat reporter who later joined NASA's public affairs staff, described NASA's handling of the *Challenger* as a "textbook case" in how not to communicate with the media.[13] What *Columbia Journalism Review* editor William Boot called the "spellbound press," dazzled by NASA's efforts to provide them access to its spectacular space feats, was replaced by scores of journalists, frustrated by NASA's lack of transparency, who did not hesitate to publish stories charging that NASA had been lax in shuttle management all along.[14]

NASA's credibility and technical judgments came under scrutiny from others it had considered to be essential supporters. According to Frank Johnson, who led NASA's public affairs office until shortly before the fateful launch, the agency's failure to be upfront from the start strongly signaled that it was not in control of the situation.[15] Whereas NASA had handled the Apollo 1 investigation on its own, President Reagan appointed an independent commission to examine what went wrong with the *Challenger* launch. The commission was chaired by former secretary of state William P. Rogers with Neil Armstrong as vice chair and included ten other leaders of the scientific, military, technical, and management communities. It identified O-ring integrity as the accident's direct, technical cause, noting that it was one of several problems that NASA was aware posed high risks to shuttle flights but did nothing to mitigate. The commissioners cited an inadequate focus on safety assurance and ineffective management and communications for dealing with technical issues as a contributing cause.[16]

The Rogers Commission also questioned the prudence of NASA's ambitious, broad-reaching sociotechnical vision and economic rationale for the shuttle. It suggested that NASA was occupied with managing payloads and increasing flight rates at the expense of safety. One commissioner, physicist Richard Feynman, accused NASA of exaggerating the shuttle's reliability, taking umbrage at the impact on people like McAuliffe. Feynman chastised the agency for seeking "to encourage ordinary citizens to fly in such a dangerous machine, as if it had attained the safety of an ordinary airliner." He insisted: "NASA owes it to the citizens from whom it asks support to be frank, honest, and informative. . . . For a successful technology, reality must take precedence over public relations, for nature cannot be fooled."[17]

Others also raised concerns about NASA's imaginary for the shuttle and its approach to realizing that vision. The House Committee on Science and Technology conducted its own investigation and, like the Rogers Commission, concluded that NASA's drive to increase the shuttle launch rate and evolve the agency into a business operation prompted it to demote its commitment to astronaut safety.[18] Senator Ernest Hollings of South Carolina rebuked NASA for launching despite expert advice. The legislator accused NASA and the Reagan administration of making the launch decision so that the president could follow through with his plan to recognize McAuliffe as the shuttle's first citizen passenger in his State of the Union speech, slated for the same day.[19] Political scientist John Logsdon wrote in *Science* that NASA's credibility was doomed from the start because the agency could not honor its claim that the shuttle would prove cost effective. A *U.S. News and World Report* article neatly summed up the situation: "The shuttle will simply never be able to provide the cheap, versatile and reliable access to space it was supposed to as the do-all and be-all of the space program . . . reality [is] that the shuttle is a complex and sophisticated vehicle—a Ferrari, not a truck."[20]

The *Challenger* accident surfaced questions about whether the nation ought to keep pursuing the shuttle program and human space flight altogether. Just two days after the incident, David Rosenbaum wrote an article for the *New York Times* titled "Should U.S. Continue to Send People into Space?" In the *Los Angeles Times*, Daniel Greenberg bemoaned the "folly of a space program fixated on humans," whom he called "the most useless cargo ever sent into orbit." Astrophysicist Thomas Gold, who had opposed the shuttle in the early 1970s, pointed out in a 1987 letter to the *New York Times* that requiring scientific, military, and commercial payloads to ride aboard the shuttle now hampered the ability to launch these assets.[21] Some citizens wrote letters to NASA and their representatives in Congress suggesting that NASA should leave the job of space flight to robotic explorers.[22]

Despite the criticisms, Americans mourned the loss of the astronauts, and most endorsed the continuation of NASA and the human space flight program. Opinion polls indicated unprecedented levels of public support for the shuttle program in the months following the accident. According to studies by Jon Miller, 97 percent of citizens regarded the shuttle as an outstanding example of American technology notwithstanding the disaster. Miller also found that the number of Americans who favored funding increases for the space program climbed by 47 percent.[23] The agency received some three hundred thousand letters, poems, and drawings from citizens in the six months following the accident, many of which conveyed the writers' and artists' belief that humans should press further into space. Young Astronaut members said that the tragedy had not discouraged but instead inspired them to want to pursue space careers.[24] Youth and adults alike showed support for restoring the nation's human space flight program by sending unsolicited monetary donations to NASA and other space-related organizations, hoping that such funds would be used to build a replacement orbiter and benefit the children of lost crewmembers.[25] Thousands of Floridians purchased special license plates commemorating the *Challenger*, and the proceeds went toward building an astronaut memorial. Kennedy Space Center set a record of hosting 2.1 million visitors the year the accident occurred even though the shuttles were grounded. John Noble Wilford wrote in the *New York Times*, "With the loss of the *Challenger* and its crew of seven, we learned, to our surprise, how much these adventures into space, into the future, mean to us as a people."[26]

Even before the accident, citizens across the nation had indicated interest in a robust space program. In 1984 President Reagan and Congress established the National Commission on Space to create a vision for the United States' next fifty years in space. Former NASA administrator Thomas O. Paine, who chaired the panel, felt that the vision should be created in consultation with ordinary Americans, recognizing public support as a critical resource for the US civilian space program. Feedback received from citizens through more than a dozen public meetings and thousands of letters conveyed to the commission that Americans wanted "a bold, imaginative civilian space effort" that included human and robotic voyages to the moon and Mars.[27]

The National Commission on Space issued its report, *Pioneering the Space Frontier*, shortly following the *Challenger* accident, but it received little attention while the Rogers Commission investigation was under way. Nevertheless, and even amid the criticism that NASA's human space flight program endured, leaders within the agency and elected officials took encouragement from the expressions of support and looked toward renewing the nation's commitment to human space exploration. James Fletcher, who returned to head NASA following the *Challenger* disaster, felt justified in continuing with human space flight, contending that "middle America" was above all interested in the astronauts. Fletcher further indicated in a February 1988 speech that public examination of the space program after the *Challenger* accident had resulted in a "consensus among people in all walks of life" that the nation needed a "major new goal in space."[28]

Accordingly, Fletcher appointed astronaut Sally Ride to chair an internal committee to help identify long-term goals for NASA. The resulting report, *Leadership and America's Future in Space*, identified human missions to the moon and Mars as ultimate goals to which the agency should strive. President Reagan included in his 1988 National Space Policy the long-range goal of "expanding human presence and activity beyond Earth orbit into the solar system" as a means of increasing commercial opportunities and asserting US leadership as the Soviet Union continued to send cosmonauts into Earth orbit. Then in 1989, on the twentieth anniversary of Apollo 11's lunar landing, President George H. W. Bush announced the Space Exploration Initiative, a plan to return humans to the moon and send them onward to Mars. Congress, however, rejected Bush's plan due to the anticipated $500 billion price tag.[29]

Harkening back to NASA's long-held view of human progression into the solar system, Fletcher still maintained that a space station would be the "next logical step" and "the key to all other steps" beyond Earth orbit that these reports and policies outlined.[30] Indeed, a space station would enable studies of the long-term effects of microgravity and space radiation on the human body and provide insights into how to build large structures in space—both of which were needed for people to stay for longer lengths of time in space. While NASA would press to gain Congress's approval of a space station over the coming years, the agency's human space flight proponents recognized that returning the shuttle to full operating status was vital to one day constructing and supplying the nascent space station. And in the near term, the vehicle was needed to fly the backlog of payloads created by the *Challenger* disaster. Because the shuttle was the centerpiece of human space flight for the time being and NASA desired to work on rebuilding its credibility and maintaining value in the public eye, the agency continued efforts to engage Americans with the space transportation system. In the approaches they took, officials negotiated the imaginary of a broadly accessible vehicle in the face of new circumstances.

Engaging the Public after *Challenger*

Following the *Challenger* accident, the agency continued to embrace some of the discursive and virtual participatory approaches to public engagement that the agency had employed during the vehicle's development and initial operating years. Communicating human space flight plans through the news media and directly with the public remained a frequently used and valuable tool for connecting with Americans and people worldwide. The agency had strived to instill confidence in the shuttle's capabilities as it began operations, and it continued to do the same in the immediate wake of the *Challenger* disaster to regain political, media, and public trust. Public affairs chief Shirley Green said in an internal memo, "My primary goals is [*sic*] to help NASA reestablish its reputation for competency as the world's leading R&D agency for aeronautics and space exploration and its credibility in the eyes of the media, the Congress and the public." The agency revised its contingency plan for shuttle accidents to clarify responsibilities and ensure that public affairs officers, shuttle managers, and other key agency personnel would work seamlessly to release known information to the media as soon as it was available and provide a statement within one hour of a mishap.[31] NASA answered citizen inquiries about the measures the agency was taking to prevent another shuttle accident, telling those who wrote in, "When we resume space shuttle flight, the spacecraft will be a safer, more dependable system, and NASA will be a better disciplined and managed agency."[32]

NASA's credibility challenges persisted even after the agency returned the shuttle to flight in September 1988. NASA was criticized in congressional hearings and the media when the orbiters suffered from a string of problems, including hydrogen leaks, processing mishaps, and main engine and booster difficulties. NASA's expenditure of $23 million to combat troubles with the shuttle's toilet also invited rebuke. Meanwhile, the agency endured media blame for botching construction of the Hubble Space Telescope's primary mirror and losing the Mars orbiter robotic spacecraft when it was about to reach the red planet. According to Marshall Space Flight Center public affairs officer June Malone, NASA public affairs officers had chosen to be "very public about how we were wrestling and dealing with those tough problems."[33] NASA held press conferences and openly answered reporters' questions to maintain transparency and avoid the speculation and distrust incurred with the *Challenger* incident, but it was difficult to shake the scrutiny.

NASA's public affairs officials continued to stress the societal relevance of its human space flight pursuits, crafting messages to resonate with various audiences. During the late 1980s, NASA issued one-page reports summarizing the results of research performed aboard the shuttle orbiters, focusing on benefits to the economy, health, and general well-being of Americans. Highlighted shuttle experiments ranged from bone metabolism research that could improve osteoporosis treatment to plant growth studies with the potential to boost terrestrial crop yields. In addition, within months of the *Challenger* disaster, the agency started the NASA Alumni League to involve former employees in restoring public appreciation of NASA by sharing information about NASA's impacts on job creation and the economy. Fletcher's successor as NASA administrator, Richard Truly, likewise engaged his leadership team in securing the assistance of the aerospace industry and space advocacy groups to promote the space program.[34]

Under NASA's next administrator, Dan Goldin, the agency continued to hear from members of Congress and the media that the agency needed to do a better job communicating its plans, achievements, and value.[35] Shortly after arriving at NASA, Goldin held public meetings in six US cities to share the agency's plans and receive feedback from citizens. NASA inferred from comments made in the meetings, which attracted more than 4,500 participants, that the public was "hungry" for information about NASA and related educational opportunities.[36] Consequently, Goldin pressed agency staff to keep up efforts to showcase the research benefits of shuttle missions. The agency issued press releases and fact sheets outlining the shuttle program's contributions to a variety of areas including airliner fuel economy, race car insulation, and artificial heart designs.[37] Like Truly, Goldin also pressured the aerospace industry to be proactive about reversing what he perceived as waning public interest in the space program. Rockwell, the orbiter's prime contractor, and United Space Alliance, a joint venture between Rockwell and Lockheed Martin to consolidate shuttle contracts into a single prime contractor, responded with newspaper ads touting the shuttle's ability to provide benefits to all people.[38]

The agency continued to associate the shuttle with not only pragmatism but also the excitement of human space flight, seeking to appeal to those who aspired to see humanity move further into space. A Kennedy Space Center brochure released after the *Challenger* accident maintained that "the original promise of the Space Shuttle remained undiminished. The Kennedy NASA/industry team was doing its part to fulfill the dream of steady progress living and working on the new frontier of space." In seeking public and congressional approval for a space station, NASA renewed the link it had originally established between the shuttle and the station. NASA officials explained to members of Congress, as they had in the past, that the shuttle was critical to enabling a station. They asserted that together the two assets would advance science and commerce in space and at the same time teach NASA how to live and work there so that humans could one day safely voyage to destinations outside of Earth's orbit.[39]

All the while, NASA also tried to rebuild public confidence in the shuttle program by once again appealing to nationalistic themes. One of the agency's principal messages in rolling out the president's 1989 budget was that the human space program had been instrumental in allowing the United States to progress and remain a great nation.[40] In a speech at Johnson Space Center just prior to the shuttle's 1988 return to flight President Reagan voiced his longstanding sentiment that the shuttle belonged to all citizens of the United States: "When we launch *Discovery,* even more than the thrust of great engines, it will be the courage of our heroes and the hopes and dreams of every American that will lift the Shuttle into the heavens."[41] During the 1990s, Goldin revised NASA's insignia once again, replacing the stylized "worm" logo with the agency's original "meatball." The move nostalgically recalled NASA's exceptional achievements in support of national greatness during the Apollo era.

Committed to giving citizens a sense of ownership of the shuttle, NASA continued to provide opportunities for them to see and virtually experience the vehicle. The *Challenger* accident prompted agency safety officials to restrict viewing locations around the Kennedy Space Center, but they nonetheless still allowed for tens of thousands of visitors to see launches from NASA property.[42] The agency established a phone hotline that provided updated launch information and continued to distribute free passes on a first-come, first-served basis for public viewers to drive onsite for launches. NASA employees could invite family and friends to attend launches and special briefings. During the mid-1990s the agency also renewed efforts to reach influential women's and minority groups, as it had done prior to the accident. According to former NASA employee Beth Beck, the agency's aim was to get them to "fall in love with NASA" and tell others about their positive experiences.[43]

In addition, NASA's visitor facilities at Kennedy Space Center and Johnson Space Center underwent major expansions during the 1980s and 1990s. Kennedy's internal newsletter explained that the center would undertake more interactive exhibits to "ensure that visitors leave [Kennedy Space Center] feeling more like owners and participants, rather than just spectators."[44] Johnson Space Center's visitor center, Space Center Houston, was designed around similar principles. Each month, Space Center Houston conducted exit interviews with one hundred visitors to understand and improve the public's experiences at the facility.[45]

NASA also continued to harness a range of communications and information technologies and techniques to make the shuttle accessible to those who could not behold it in person. NASA Select TV, on US cable television, provided twenty-four-hour coverage of certain shuttle missions to millions of homes and schools.[46] The channel broadcast the Hubble Space Telescope repair mission in 1993 and subsequent servicing efforts. The agency affixed high-resolution cameras to the shuttle's external tank to give television (and later, internet) viewers a new perspective on the vehicle's launch and ascent into space. During the 1990s NASA's public affairs office secured more on-orbit mission time for press conferences and interaction with media and schools.[47] NASA's education and public outreach proponents tapped into pop culture in teaming with Disney to send a Buzz Lightyear action figure from Disney-Pixar's *Toy Story* to the International Space Station aboard STS-124 in 2008 to promote a set of jointly developed online educational video games featuring the character.[48]

The internet's emergence in the mid-1990s generated a new excitement within NASA about the possibilities of engaging directly with the public, and the agency became an early adopter of website development for delivering up-to-date information about shuttle missions. Goldin's dedicated page on the NASA website stated that the agency was "deeply committed to spreading the unique knowledge" that came from its research and saw the internet as a promising way "to expand this outreach exponentially."[49] STS-71, which launched on June 27, 1995, became the first shuttle mission to have its own web page, while NASA's new "Shuttle Web" service provided real-time mission updates. The October 1998 STS-95 mission that flew John Glenn, an astronaut from the Mercury missions who became a US senator, was the first to include a live webcast from space. Johnson Space Center answered public questions about the shuttle program through an online forum called "Ask the MCC" (the abbreviation for Mission Control Center) for several years beginning in the late 1990s.[50]

NASA officials also continued to engage schoolchildren with human space flight. Their commitment aligned with the Reagan and George H. W. Bush administrations' focus on American education reform after multiple reports linked national global competitiveness to the preparedness of the workforce to excel in research and technological innovation. NASA's education office also yearned to fulfill McAuliffe's aim to connect closely with the K–12 education community. After receiving authority from Congress in 1987 to replace *Challenger* with a new orbiter, the agency invited America's elementary and secondary students to participate in a competition to name the vehicle. In May 1989 President Bush announced that the new shuttle would be known as *Endeavour*. While many schools had proposed the name, NASA attributed its decision to Tallulah Falls School in Georgia and Senatobia Falls Middle School in Mississippi, whose project entries scored highest.[51] NASA administrators Truly and Goldin each promoted still other initiatives connecting the shuttle to students and educators. These included using public donations made to NASA after the *Challenger* disaster to establish a fund for teacher scholarships, bringing underserved students and teachers to shuttle launches, continuing ham radio communications between schools and shuttle (and eventually International Space Station) astronauts, and having astronauts teach science lessons from orbit.[52]

Revising Shuttle Use Policy

NASA's commitment to engaging with the public through discourse and visual and virtual means remained largely unchanged in the post-*Challenger* era. However, concerns about shuttle safety and reliability threatened the agency's carefully laid vision for the shuttle as a "space truck" intended to haul cargo, retrieve and repair damaged satellites, and conduct experiments for a variety of customers. Pundits and politicians asked whether human-tended vehicles were necessary to carry payloads. Couldn't an expendable vehicle just as easily have hoisted the Hughes communications satellite that rode in *Challenger*'s cargo bay? Moreover, NASA's two-and-a-half-year effort to return the shuttle to flight created a massive backlog of probes awaiting rides to space, delaying replacement of defense and communications satellites and slowing the deployment of major telescopes on which space scientists' work depended. The Rogers Commission advocated that the nation revisit its policy to depend on a single launch capability.[53] Four years later, a committee established by Vice President Dan Quayle and chaired by Norman Augustine of the Martin Marietta Corporation to provide advice on the space program's future echoed these sentiments. The Augustine committee concluded that relying on the shuttle to achieve what expendable vehicles could do inappropriately jeopardized human lives and the shuttle orbiters. Investing in new heavy-lift expendable vehicles, the committee asserted, would assure access to space while lessening schedule pressure and risk associated with flying the shuttle.[54]

The vise on the vision of a broadly accessible shuttle grew as political figures reshaped space transportation policy. Six months after the *Challenger* accident, Congressmen Bill Nelson and Bob Walker issued a bipartisan call "to develop a strong and definite policy for access to space that includes maximum flexibility and efficiency." They pushed to build a replacement shuttle orbiter, acquire expendable launchers to work through the backlog of shuttle payloads, and allow private satellites to fly on either commercial launch vehicles or the shuttle. NASA leadership had concurred with the Rogers Commission's recommendation to move to using a mixed fleet of vehicles to satisfy launch requirements and, like the congressmen, saw the shuttle as continuing to play a part in serving commercial access to space. Over NASA administrator Fletcher's objections, however, Reagan released a statement in August 1986 that stepped back from regarding the shuttle as a market-focused enterprise. Reagan continued to venerate the shuttle: he pushed to build an orbiter to take *Challenger*'s place, stating that "NASA and our shuttles will continue to lead the way, breaking new ground, pioneering new technology, and pushing back the frontiers." But the president also sympathized with the nascent commercial launch industry, which saw the shuttle as a competitive threat. Reagan therefore indicated that "NASA will no longer be in the business of launching private satellites"—a task, he said, that could be done "better and cheaper in the private sector." A few months later, the president codified the decision in a policy directive, stating that the shuttle would launch commercial or foreign payloads only if its unique capabilities were required or there were compelling national security or foreign policy reasons to do so. The policy also affirmed that US government payloads would rely on shuttles as well as expendable vehicles.[55]

NASA had sold the shuttle to the nation on the premise that it would serve a variety of customers. Reagan's policy shift forced the shuttle's main purpose to slip away where most major payloads were concerned. The agency continued to serve scientists by launching probes that had been designed to fit within the orbiter's payload bay and were too large to fit on any other vehicle, such as the Hubble Space Telescope, Compton Gamma Ray Observatory, and Jupiter-bound *Galileo* spacecraft. But despite the mixed-fleet provision, future major payloads would not fly on the shuttle. Fletcher cancelled the agency's effort to develop an upper stage to boost spacecraft to the proper orbits and trajectories because of the risk of carrying rockets with highly combustible liquid fuel aboard the shuttle. Meanwhile, the air force terminated plans to commission a shuttle launch site in California to put payloads into polar orbits, and

Congress authorized the military to purchase dozens of Titan IV heavy-lift expendable launchers and smaller rockets to meet its needs.[56]

Looking back, former administrator James Beggs lamented, "The military abandoned us, and commercial payloads were taken off, and the [shuttle] program kind of went into the doldrums."[57] Beggs especially rued the Reagan administration's decision to block making the shuttle available for commercial use, in part because it gave much of the world's heavy-lift launch market to France's Ariane rocket and also because it impinged on NASA's ability to preserve its democratic philosophy regarding the shuttle.[58] Shuttle customers had helped to give the vehicle purpose and legitimacy. Now the agency could not depend as extensively on an array of users of the vehicle to substantiate its value. Beggs considered the reversal of the shuttle user policy "one of my regrets after the *Challenger* [accident]."[59]

Another major challenge to NASA's ability to realize wide use of the shuttle was the fact that the vehicle would never launch dozens of missions per year as it had once claimed. Eighteen months following the accident, NASA projected flying fewer than one-third of the missions previously planned through 1992. Needing to establish payload priorities, the agency revoked its commitment to flights of nonscientific cargo, such as artwork. NASA announced its termination of the Nonscientific Payload Program in the *Federal Register* without inviting public comment, simply stating that the program "has served its purpose and is no longer in keeping with current policy."[60] The agency also significantly changed the Shuttle Student Involvement Program. While NASA eventually flew the experiments of students who had won competitions prior to the *Challenger* accident, officials felt they would be unable to keep up with flying subsequently selected experiments in a timely fashion. Instead, NASA rebranded the initiative as the "Space Science Student Involvement Program" and presented it as a competition for high school and middle school students to propose experiments that *could someday* be performed in space. The agency awarded internships, scholarships, trips to Space Camp, and computers but not shuttle experiment flights. NASA later expanded the program to include elementary students and held competitions for news articles, art, and spacecraft design to promote scientific literacy and creative thinking among students with a range of interests.[61]

Just as the shuttle's reduced availability prompted NASA to make choices about the vehicle's use, this factor also drove private companies to reconsider whether to leverage the shuttle to perform experiments. Prior to the accident, commercial business had not materialized at the level NASA had anticipated due to flight costs and scheduling challenges; the ensuing backlog of payloads only exacerbated the limitations on shuttle access and discouraged private sector flight commitments. According to McDonnell Douglas payload specialist Charles Walker, once commercial payloads took lower priority after the *Challenger* accident and it appeared that flights would not reach levels NASA had projected, his company's management cancelled plans for electrophoresis as well as subsequent experiments. Former NASA human space flight manager Michael Hawes pointed out that many other corporations also reassessed their interest in performing research aboard the shuttle.[62] The slump in commercial demand had ripple effects on other sorts of private involvement with the vehicle. For example, the company Spacehab, whose cargo bay module contained dozens of lockers like the ones contained on the shuttle's middeck to create more space for flight experiments, had a difficult time attracting commercial interest and ended up supplying carriers to NASA to fly agency-funded experiments.

Even so, the agency remained determined to attract and serve customers in government, commercial, and university domains interested in using the shuttle as a platform for space-based research. The European Space Agency's Spacelab module flew more than two dozen times through 1998, and its unpressurized pallets flew as late as 2008. NASA continued to award grants to scientists to fly experiments in the shuttle's middeck lockers. The agency also kept its commitment to the popular Get Away Special (GAS) program, desiring to provide low-cost access to space for a variety of researchers. Interest remained high among private individuals, small companies, and other organizations in flying modest scientific payloads aboard the shuttle, and reservations to purchase GAS cans poured in even after the *Challenger* disaster. However, with a formidable backlog of primary payloads and nearly five hundred GAS experiments awaiting launch, NASA announced in 1987 that it was closing the reservation queue to new payloads until after the shuttle returned to flight. It resumed GAS experiment flights in 1991 and accommodated 114 payloads over the next eleven years. When it reopened the reservations list in 1992, it was much shorter, as many would-be GAS investigators had withdrawn, unable to plan for flight in the wake of the *Challenger* accident.[63]

The GAS program nonetheless endured some policy changes. With safety concerns heightened following the shuttle accident, NASA reinstated its restriction on the ejection of small satellites from GAS canisters. Those who wanted to deploy payloads would have to fly small satellites on a Hitchhiker carrier or a "complex autonomous payload," which could cost up to ten times as much a GAS can. In addition, NASA raised prices for most users to adjust for inflation since the program's inception in the mid-1970s. Whereas a full-sized canister formerly cost $10,000, the price for the same container climbed to $27,000.[64] The new prices, however, were still far below the true costs borne by NASA.

As a shuttle user community, educators and students again continued to receive special attention from NASA. After educators protested the GAS can price hikes as prohibitive, the agency decided to waive the higher fees if students were involved throughout the entirety of a project.[65] Beginning in 1996, the agency further accommodated student shuttle experimenters by offering a new initiative, the Space Experiment Module (SEM) Program. Student teams from kindergarten through college could apply to fly experiments at no cost inside NASA-provided GAS canisters using the new modules, which took care of power and control requirements. The SEMs allowed students to focus on creating their experiments rather than on managing the operational infrastructure. With the changes to the Shuttle Student Involvement Program, SEMs became the primary means for students to fly their own experiments in the post-*Challenger* era.

The shuttle also accommodated Hitchhiker payloads developed by educational organizations. One project, Starshine, resulted from a partnership among the Utah Department of Education, the Hansen Planetarium in Salt Lake City, amateur astronomers, and amateur radio enthusiasts. The partners built a trio of satellites designed to be observable from Earth so that students could measure their movements. Two satellites were launched by the shuttle, on STS-96 in 1999 and STS-108 in 2001; a third satellite was launched by an expendable launch vehicle (see fig. 7.3).[66] Some 120,000 students in forty-three countries participated.[67]

FIGURE 7.3. Phyllis Moore, codirector of Project Starshine, stands inside the payload bay of space shuttle *Endeavour* with the *Starshine-2* satellite ahead of the launch of STS-108 in 2001. Photograph from NASA.

Meanwhile, Ride and colleagues at NASA's Jet Propulsion Laboratory and the University of California, San Diego proposed putting an Earth-facing camera on the shuttle that could be controlled by middle school students in their classrooms. According to Ride, the project, dubbed KidSat, allowed kids to "feel like they were participating in a very real way."[68] Students would write research proposals for images to be taken by an electronic still camera, which was connected to a laptop computer aboard the shuttle. Undergraduate students at UCSD would develop and upload code directing the camera where to point and then send the middle schoolers the photos over the internet for analysis. The camera flew on six shuttle missions starting in 1996 before moving to the International Space Station and becoming known as ISS EarthKam. The project involved seventy-three thousand students in seventeen countries through 2012.[69]

While NASA continued to facilitate shuttle use as best it could given the policy and flight rate constraints following the *Challenger* accident, by the mid-1990s finding customers for the shuttle was no longer quite so paramount for the agency. After years of making the case for its next step in human space flight, NASA secured congressional approval in 1993 to develop a space station, and the shuttle would play an integral role in its assembly. First, though, NASA flew a series of shuttle missions to the Russian space station *Mir*. They were a technological and diplomatic prelude to collaborating with the United States' former foe on what would ultimately become not just an American space research laboratory but a global one, to be dubbed the International Space Station. Then, beginning in 1998, the shuttle assumed the role that the agency had articulated for it some three decades earlier: hauling elements of the station for assembly in orbit two hundred miles above Earth.

NASA's search for customers would eventually turn to the station. Officials envisioned that the station would provide a stable, longer-lasting platform for experiments that would theoretically produce even more meaningful science than the shuttle allowed. Just as it did for the shuttle, NASA sought to create a community of station researchers and worked to generate awareness among academic institutions and private companies of the forthcoming facility and its value for life and microgravity sciences studies and manufacturing purposes.[70] In the meantime, the agency continued to accept customer experiments to fly on the shuttle. However, given the monumental demands of assembling the station, the shuttle had little room—and astronauts had little time—to accommodate the payloads. Devoting the shuttle to station construction created a backlog of small research payloads, including GAS experiments. Meanwhile, microgravity investigations with commercial potential awaited flight aboard the not-yet-completed International Space Station. NASA responded by developing STS-107, a shuttle mission that would be dedicated to carrying dozens of research experiments using the Spacehab module. The mission would use the shuttle *Columbia*, which was too heavy to loft station components.

The issue of welcoming paying users and uses of the shuttle did not go away altogether with the station's arrival. Recognizing that the agency could not accept the range of customers it could prior to the *Challenger* accident, some human space flight advocates started thinking ahead to how to reduce shuttle operating costs after completing the International Space Station. In 1997 United Space Alliance publicly announced its desire to market space for rent aboard *Columbia*. The company wanted to fly privately owned satellites as well as commercial advertisements aboard the vehicle for a fee.[71] The Russians had been securing corporate sponsorships and displaying companies' names and logos aboard their own *Mir* space station to keep the spacecraft solvent; members of Congress, including Representative Dana Rohrabacher of California, and space advocacy groups such as the National Space Society (formed from the merger of the National Space Institute and the L5 Society in 1987) and the Space Frontier Foundation backed using the approach to minimize shuttle and station costs. NASA's Office of Space Flight supported allowing a broader range of payloads on the shuttle and possibly even fully commercializing at least one orbiter.[72]

Proponents of widening shuttle use in these ways met with partial success. The Commercial Space Act, passed in 1998, established commercial use as a primary goal of the International Space Station while directing NASA to conduct a study about the feasibility of privatizing the shuttle orbiters.[73] The agency allowed United Space Alliance to solicit payloads from private customers for two shuttle missions. In 2000 it formed a partnership with a company called Dreamtime Holdings to put high-definition cameras on the shuttle and station and to develop new multimedia products and programming featuring these spacecraft and other NASA activities.[74] But the Commercial Space Act also limited the agency's ability and willingness to engage in pursuits that could compete with the private sector. That same year, a company that went into business flying the ashes of cremated individuals lodged a complaint with NASA about unfair competition after discovering that the agency had decided to allow the cremated remains of a renowned artist to be flown on the shuttle at no cost. NASA consequently drafted a policy banning shuttle flights of any other human or animal remains.[75] The agency also ruled against allowing advertising on the outside of the shuttle or station.[76] NASA's quest to invite new shuttle program participants and shore up the vehicle's financial posture remained at odds with not encroaching on commercial economic aims or adorning a taxpayer-funded spacecraft with corporate logos. How to serve and balance public and private interests in the shuttle never ceased to be a struggle for the agency.

Reevaluating Shuttle Flyer Policy

One of the most palpable ways in which NASA had sought to realize the vision of making the shuttle program more relevant and accessible to Americans was to diversify the vehicle's crews. The agency had expanded eligibility beyond the realm of military test pilots, actively recruited women and people from racial and ethnic minorities, invited corporations to fly their own personnel with their payloads, and begun to welcome people with different professions and types of experience aboard. The STS-51L *Challenger* crew had reflected these changes.

After the accident, selecting career astronauts continued in the same vein. In 1988 NASA began to recruit astronauts on a two-year cycle and welcomed classes ranging from fifteen to thirty-five new pilots and mission specialists at a time.[77] The agency remained committed to gender, racial, and ethnic diversity in its hiring. Many more women and people of color were selected as astronauts, leading to additional firsts-in-flight in the post-*Challenger* era. Physician Mae Jemison became the first Black woman in space aboard STS-47 in 1992, while Ellen Ochoa made history as the first female Hispanic astronaut to fly into space on STS-56 in 1993. A Native American astronaut, navy pilot and registered member of the Chickasaw Nation John Herrington, first flew on STS-113 in 2002. Eileen Collins broke more than three decades of history of male-piloted American space missions when she piloted STS-63, a shuttle mission to the Russian *Mir* space station, and then became the first female shuttle commander on STS-93 in 1999.

The *Challenger* disaster nonetheless had the effect of tempering NASA's ambitions to further widen the scope of those it might fly. Before the accident, enthusiasm for broadening the ranks of shuttle flyers emanated from the Reagan administration, members of Congress, the media, and citizens. Supportive entities saw space as the domain of everyone and the shuttle as the nation's equal opportunity provider. NASA had safely flown nearly two dozen payload specialists—trained less extensively than career astronauts—during the first five years of shuttle missions. But the loss of *Challenger*, carrying teacher Christa McAuliffe as well as Gregory Jarvis of the Hughes Corporation, prompted NASA to question the wisdom of doing so. Former shuttle program manager Wayne Hale admitted, "It was pretty clear that was not the kind of vehicle we had technologically."[78]

Various civic and educational organizations expressed their desire to see NASA continue to include noncareer space flyers in the tragedy's aftermath. McAuliffe's presence aboard *Challenger* had sparked a great deal of interest in the mission among Americans. That so many children and other citizens wrote to NASA imploring the agency not to end the Teacher in Space Program indicates the wide support that existed for flying people of all walks of life despite the tragedy. Public speaking requests to McAuliffe's backup, Barbara Morgan, and the other Teacher in Space Program national and state finalists soared. The US Department of Education, National Education Association, National Science Teachers Association, and Young Astronaut Council all urged continuation of the Teacher in Space program.[79]

In the weeks following the accident, President Reagan and NASA expressed their resolve to fulfill the dreams so many had for the *Challenger* flight and its promise of democratizing human space flight. Speaking on national television after the accident, Reagan mourned America's loss while stating affirmatively: "There will be more Shuttle flights and more Shuttle crews and yes, more volunteers, more civilians, more teachers in space."[80] When Reagan visited an Alexandria, Virginia, high school and a student asked about plans to fly another teacher, the president stated, "I don't believe that this tragedy should in any way affect the policy that we had," noting that space exploration "should not just be left to scientists or career people." The president added: "The space program belongs to all of us and to the people."[81] Two weeks after the accident, NASA's acting administrator, William Graham, affirmed the agency's continued support for the Space Flight Participant Program and the Teacher in Space initiative. Graham announced that Morgan would be offered the opportunity to fly as the next citizen aboard a shuttle mission and also stated that NASA still intended to fly a journalist.[82]

But as the Rogers Commission's investigation unfolded, revealing the shuttle to be more experimental than operational in nature, opposition to flying citizens not trained as rigorously as NASA's astronauts emerged among some scientists, engineers, astronauts, and members of Congress. John G. Stewart, a member of NASA's Aerospace Safety and Advisory Panel, was "ashamed" that the standing panel had not raised questions about citizen flights sooner. Senator John Glenn disputed putting a "lay person" in space for what seemed to him to be the purpose of "gaining public support" while the shuttle remained relatively new. Senator Hollings, ranking member of the Senate Commerce Committee that had jurisdiction over NASA, had supported the Space Flight Participant Program. However, he thought that, for the time being, shuttle flights should be left to "the professionals."[83] Veteran shuttle flyers John Young, Robert Crippen, and Richard Truly believed the door should not be closed forever to citizen flights but that NASA first needed to gain a better understanding of the vehicle.[84]

As internal debates about the future of the citizens-in-space effort swirled, officials in NASA's education and public affairs organizations argued that NASA would risk its credibility if the agency did *not* continue with the initiative. Educational affairs director Robert Brown contended that withdrawing would "be a demoralizing blow to the motivational and aspirational space spirit that the space program has generated among the Nation's school children and adults alike." Green concurred, noting that "to reverse course now would seem to confirm the accusations" that the program was just a public relations stunt.[85] But the desire to restore the agency's image of competence and responsibility weighed heavily on Fletcher, who had returned to the NASA administrator post in 1986. Fletcher prioritized proving the shuttle's reliability and safety and focusing on projects that required the vehicle's unique capabilities over reestablishing citizen flights. He did commit NASA to assessing annually when it could safely assign Morgan to a mission and open flights to others. However, NASA suspended the process to select a journalist as a Space Flight Participant and informed congressional leadership that it would defer initiating its feasibility study on flight opportunities for disabled persons until learning more about shuttle safety following the *Challenger* accident.[86]

Further, NASA issued a new policy in the *Federal Register* in 1989, acknowledging that "the Challenger accident marked a major change in the U.S. outlook and policies with respect to the flight of other than NASA astronauts" and prompted the agency to "re-examine previous understandings, expectations, and commitments regarding flight opportunities."[87] According to the new rule, NASA would look to its mission specialists to perform flight tasks and would fly people outside of career astronauts only if their presence was necessary to perform specialized functions. The new policy allowed for foreign payload specialists but, together with the Reagan administration's decision to remove payloads not requiring the shuttle's unique capabilities, led the corporate payload specialist program to wither. The rule did, however, make room to provide flight opportunities to people deemed by NASA to "contribute to other approved NASA objectives or to be in the national interest." The policy considered a teacher in space as falling within this category.

In 1990, the Soviet Union announced plans to fly a Japanese journalist to its orbiting space station, *Mir*, aboard a *Soyuz* rocket. Commercializing its space program to bring in much-needed cash, the former anticapitalist superpower signed a $12 million contract with the Tokyo Broadcasting System for reporter Toyohiro Akiyama's December 1990 trip.[88] In contrast, NASA approached the fifth anniversary of the *Challenger* accident without having held any reviews regarding Morgan's flight. By then Truly had succeeded Fletcher as NASA administrator. After the National Science Teachers Association and other education organizations raised concerns to President George H. W. Bush about the Teacher in Space program's future, Truly convened an internal advisory group to look into the matter. The following year Truly left his post due to differences with the Bush administration, but days before he departed he concluded that the time had come to assign Morgan to a flight and declared: "She's ready, the Space Shuttle is ready, and the American people are ready for the educational inspiration that flying Barbara will provide."[89]

Goldin, the incoming administrator, was unsure about flying Morgan. Just three weeks after Goldin's arrival at NASA, the House Committee on Science, Space, and Technology challenged Truly's assessment, stating in a report accompanying NASA's authorization bill that the risks made putting a teacher on the shuttle "highly unwise." Goldin needed Congress's support to allow NASA to proceed with the space station, whose projected costs were growing considerably. The agency also was dealing with bad press about the Hubble Space Telescope's flawed primary mirror and hydrogen leaks on two shuttle orbiters. The new administrator noted in a 1993 speech that "for the NASA of today, the specific challenge is to win back our credibility through performance." For Goldin, that meant not only completing projects within budget but also upholding the agency's shuttle safety record since the 1988 return to flight. The NASA chief was nervous that the agency had more to lose than to gain with Congress by resuming the citizens-in-space program.[90] Goldin maintained publicly that flying noncareer astronauts was too risky.

Pressure to fly Morgan, however, continued to come from many sectors. Representative Larry LaRocco and Senator Dirk Kempthorne urged Goldin to reconsider.[91] Teachers deluged NASA and the White House with letters and petitions, and NASA's education office also continued to staunchly support her mission.[92] Morgan herself challenged Goldin's claim that shuttle flights were too risky for teachers, telling the administrator in a letter that teachers accepted physical risks every day in going to their classrooms and that astronauts were no more expendable than teachers. Goldin admitted privately to Morgan his personal wish for her to fly but informed her in June 1994 that the "highly visible nature of the program and the diversity of opinion on whether it is appropriate to allow a teacher to accept the risks illustrates the need to consider all views."[93]

As Truly had, Goldin directed an internal review of the matter. And again a group of representatives from various NASA organizations noted the potential boon that a successful mission including a teacher could have for the nation's science education goals and public awareness of NASA. The reviewers also cautioned that a second accident could lead to "the demise of NASA," but they nonetheless validated education as a NASA mission and endorsed selecting science and math teachers as shuttle mission specialists, with Morgan to be the first of this new class of astronauts.[94]

Before Goldin settled the issue of flying Morgan, he found himself presented with another special flyer to consider. Mercury astronaut John

Glenn had long desired to return to space. Now a septuagenarian serving in the US Senate, he proposed to Goldin that he could participate in a flight to contribute to research on muscle and bone loss and sleep pattern changes in older people. As Goldin and senior NASA officials contemplated the prospect, media reports broke the news that Glenn might fly on the shuttle. "It would be good for America," said John Pike of the Federation of American Scientists, recognizing Glenn as a national hero whose flight aboard the shuttle could cast positive attention on the space program. President Bill Clinton publicly endorsed the idea.[95]

At the same time, there were numerous arguments against a flight with Glenn aboard. NASA had no plans to send older people into space and encouraged astronauts to retire before reaching anywhere near Glenn's age. And if the intent truly was to aid ground-bound senior citizens, then wouldn't it make more sense to invest instead in more geriatric research on Earth with a large number of volunteers rather than a single data point from space? Some journalists speculated that Goldin was entertaining flying Glenn as a political favor to President Clinton, whom Glenn had defended during Senate hearings on campaign finance abuses. The Space Frontier Foundation argued that flying Glenn would be a slap in the face to career astronauts and American taxpayers when NASA had pulled back on broadening citizen access to shuttle flights. The National Space Society linked the question of flying Glenn to that of launching Morgan, stating that NASA should not send Glenn without also allowing Morgan to go to space and restarting the citizens-in-space program.[96]

Goldin took these views into advisement, and on January 16, 1998, NASA publicly announced plans to assign both Morgan and Glenn on upcoming shuttle flights. Goldin gave a speech confirming the agency's plans to fly Glenn as a payload specialist along with the crew of STS-95, a dedicated scientific research mission scheduled for the fall that would conduct eighty-eight experiments. His speech lauded Glenn's willingness to accept the risk of returning to space to serve the country and benefit the lives of older Americans. But he made no mention of Morgan. Instead, NASA shared news of her selection as a mission specialist in a press release. The unceremonious release stated that NASA had "determined that it is appropriate to include educator mission specialists" among its astronaut ranks, although it remained silent about a flight assignment.[97]

Throughout history, NASA administrators had sought to leverage the agency's ability to elevate student interest in science and engineering careers, but Goldin's successor, Sean O'Keefe, took that commitment to a new level. O'Keefe, who had served as deputy director of the Office of Management and Budget as well as a professor of business at both Penn State and Syracuse University, made education a pillar of his vision for NASA, adding the aim "to inspire the next generation of explorers" to the agency's new mission statement. Bolstering this focus, President George W. Bush signed the No Child Left Behind Act into law during the first weeks of O'Keefe's tenure. Consequently, O'Keefe was primed to fulfill McAuliffe's legacy and finally put another teacher on board the shuttle. In an April 2002 speech at Syracuse University, O'Keefe announced that Morgan would "begin her mission as the first Educator Mission Specialist."[98] Eight months later NASA assigned her to STS-118, a mission to the International Space Station scheduled for November 2003.

O'Keefe was determined that Morgan would not be the last educator to fly as a mission specialist. Under his leadership, NASA established the Educator Astronaut Program, which would capitalize on teachers' unique capabilities to inspire students. Candidates would need to be certified K–12 educators with a minimum of three years of teaching experience and at least eighteen credit hours of science, math, or engineering. NASA rolled out a website inviting students and the public to nominate accomplished teachers. The agency ran televised public service announcements for the program in English and Spanish in an effort to attain a diverse pool of applicants.[99]

Just ten days into the recruitment period, however, tragedy struck the shuttle program again when STS-107/*Columbia* disintegrated on its February 1, 2003, return to Kennedy Space Center. Once more, the nation mourned while the agency grappled with the future of the human space flight program. As with *Challenger*, citizens wrote to NASA with encouragement to keep alive the commitment to include educators on shuttle missions, and applications continued to roll in after the accident. Whereas the Space Flight Participant Program was paused indefinitely after the *Challenger* disaster, the Educator Astronaut Program received

FIGURE 7.4. Educator astronaut and mission specialist Dorothy Metcalf-Lindenburger aboard space shuttle *Discovery* on STS-131. Photograph from NASA.

the go-ahead from NASA leadership to press onward following the *Columbia* accident. The agency received nearly 1,700 applications, and program managers forwarded those with superior qualifications to Johnson Space Center for review with all other applications for NASA's 2004 astronaut class. Of the eleven astronauts NASA selected, three were educator-astronauts.[100]

After the shuttle mission manifest underwent modifications following the *Columbia* tragedy, Barbara Morgan finally flew in 2007. Many of the original Teacher in Space finalists were onsite at Kennedy Space Center to witness the launch, more than twenty years after the quest to put an educator on the shuttle began. Two of the new educator-astronauts, Joseph Acaba and Ricky Arnold, were crewmembers of STS-119, a shuttle mission to the International Space Station in 2009. The third educator-astronaut, Dorothy Metcalf-Lindenburger, flew to the station aboard STS-131 in 2010 (see fig. 7.4).

Some entities envisioned allowing even broader access to the shuttle. For example, just as United Space Alliance considered offsetting shuttle operations costs by making *Columbia* available to fly commercial payloads, the company also gave thought to opening the vehicle to private paying passengers.[101] But NASA no longer saw a value proposition for using the shuttle to further democratize space flight. While the agency's plans to fly Glenn and Morgan seemed to signal its rising confidence in shuttle safety and a desire to resume expanding access to space, Goldin asserted that space flight remained risky and that NASA would not fly civilians with only brief training. In addition, the agency most needed to fly highly trained people who could construct the station. Given those priorities, seats on the shuttle went to career astronauts and occasional payload specialists.[102]

These limitations were stretched, however, when Dennis Tito, a former NASA engineer who went on to become a multimillionaire businessman, paid the Russian Space Agency $20 million to fly to the International Space Station. The Russians had been flying paying passengers to *Mir* and had worked a deal to send Tito. But Russia could not afford to support *Mir* and also contribute to the International Space Station. After deorbiting *Mir* in early 2001, the Russians invited Tito to visit the International Space Station instead. Neither NASA nor the other station partners had the power to stop the Russians; partners were required to notify each other about whom they would fly to the station but did not have to ask permission. Despite advocating commercial activity in space, Goldin protested Russia's plan. The NASA administrator worried that Tito's presence on the station, which was still under construction, would create a distraction and safety threat to the astronauts working on assembly. Goldin was joined in his objections by key congressional stakeholders. Representative Ralph Hall, for one, did not like the idea of prioritizing private citizens who were willing to pay to fly into space when NASA had "an obligation to the American taxpayers to ensure that the International Space Station is not jeopardized during its assembly or its resources misallocated." Despite the controversy around his own recent flight aboard the shuttle, Glenn added that allowing Tito to fly trivialized the costly space station.[103]

Tito's voyage nevertheless went forward in April 2001, without incident. Afterward, Goldin publicly claimed that NASA did not oppose trips by noncareer astronauts once the station was ready. NASA's Hawes explained that the issue was timing: "We weren't trying to just give negative answers because we didn't want to fly other people. We just got overwhelmed with the constraints of [ISS assembly]." Following Tito's flight, the station partners created guidelines for visits by noncareer astronauts. Even so, NASA never endorsed Russia's subsequent flights of paying passengers—a stance Alan Ladwig lamented as a missed opportunity. "NASA was really almost anti–Tito's flight," Ladwig said. "And I never fully understood why because we're having a hard time selling the space program, and here you have a guy who's willing to spend his own money to go to [space]. I've never understood why we didn't embrace that."[104]

Tito returned from his flight as an advocate of opening space travel to a broader range of people. He told *Today*'s Katie Couric: "I would like to work with NASA to encourage them to fly ordinary people at government expense on the seventh seat of the shuttle so that we can get a cross-section of people . . . to experience what it's like up there."[105] He testified at a congressional hearing on space tourism that "we need to find ways to include the general public in our human space flight activities" and argued that NASA should reinstitute the Space Flight Participant Program. "We need to once again offer our nation's teachers, journalists, creative artists and others an opportunity to experience what is now the sole bailiwick of fighter or test pilots and scientists," Tito said. "The bottom line is that the American people, who pay for the space program, should have every opportunity to share in it.[106]

The hearing also considered whether NASA might expand flight opportunities by following Russia's example and selling shuttle seats. It was an idea staunchly supported by Congressman Rohrabacher. But just as NASA hesitated to expand commercial uses of the shuttle and station without limit, so too did the agency resist broadening access to space flight when private entities were attempting to forge a space tourism industry. In the mid-1990s, the independent XPRIZE foundation had announced a competition seeking the first privately built spaceship capable of sending humans to space for a prize purse of $10 million. The offering had spurred several enterprising teams from around the world to pursue this goal and start space tourism businesses. Testifying at the space tourism hearing alongside Tito, Hawes stated that NASA was committed to "opening up the space frontier for commercial purposes including tourism" while avoiding competition with the private sector.[107] By the turn of the millennium, NASA officials still believed in making space flight accessible to more people. That conviction, however, was not tied as strongly to the shuttle as it had once been now that the private sector was actively pursuing this vision.

8

Commemorating the People's Spaceship

Shuttle program developers anticipated that each orbiter would be able to fly a hundred missions with appropriate maintenance. NASA undertook major upgrades to address problems experienced during flights in the early 1980s and then after the *Challenger* accident. The agency approved additional enhancements to allow the shuttle to reach the *Mir* space station as well as the International Space Station. After a brief "design freeze," NASA allocated funds only for minor upgrades to support missions, improve safety, and reduce costs and obsolescence. The agency targeted more substantial resources toward developing reusable launch vehicle concepts for safer, more capable, and more efficient transport of humans and cargo to space.[1] NASA administrators Dan Goldin and Sean O'Keefe both strived to keep the shuttle going by pursuing upgrades to the vehicle as well as initiatives to replace or complement the shuttle and help extend its life to as late as 2020.

NASA's timeline to make decisions about the future of the space shuttle was hastened on February 1, 2003. That morning, *Columbia* and the STS-107 mission crew were lost when the orbiter broke apart as it streaked across the skies above the lower central United States, just minutes from its scheduled return to Kennedy Space Center. The timing and technical causes of the accident differed from the *Challenger* experience. During *Columbia*'s launch, a piece of foam from the external tank had broken loose and struck the thermal protective tiles on the leading edge of the orbiter's left wing. The damaged wing was unable to deflect the searing heat generated when the orbiter reentered Earth's atmosphere, which caused the vehicle to disintegrate. What the accidents had in common was that they both were indirectly the consequence of pressure to keep a highly complex technological system operating on schedule with a limited budget. Only a few years earlier, in 1998, NASA's Aerospace Safety Advisory Panel had cautioned that the agency was not investing adequately in shuttle safety. The following year, the panel pointed out that *Columbia* had recently experienced wiring shorts and hydrogen leaks that served as "a harbinger of things to come" as the orbiters aged.[2]

For NASA and the George W. Bush administration, the *Columbia* accident provided the impetus to bring the shuttle program to a close and to reimagine the nation's human space flight program once again. The new focus, Bush announced, would be on extending human presence throughout the solar system, beginning with exploration of the moon in preparation for travels to Mars and beyond. Partnerships with industry, academia, and international space agencies would be necessary, but the vision did not involve access to space for more Americans as the shuttle had.

Because NASA needed the shuttle to complete space station assembly and could not afford to begin the new human exploration program in earnest until it retired the orbiters, the agency would keep the vehicles going until the end of the decade. In its twilight years the shuttle no longer served its original purpose of directly meeting many users' needs. However, NASA worked to keep the shuttle in the public eye until the program's end, creating ways to allow Americans to engage with and celebrate the vehicle that the agency had originally branded as a spaceship belonging to all.

According to Michael Cabbage, who was a space beat reporter at the *Orlando Sentinel* when the *Columbia* accident occurred, the catastrophe did not feel quite like the *Challenger* episode. Unlike *Challenger*, the event was not captured on live television. It did not involve the flight of an "ordinary" citizen like schoolteacher Christa McAuliffe.[3] And the nation had already come to face the reality of a shuttle disaster almost exactly seventeen years prior. It was no less tragic of an event than *Challenger* had been, but the *Columbia* incident seemed less jolting and felt to many Americans, Cabbage said, like a "giant airline accident."[4]

The public also experienced the *Columbia* disaster differently because this time NASA was much more transparent than it had been following the *Challenger* tragedy. Indeed, the agency had learned painfully from that incident how critical open communication was to its credibility. As soon as NASA officials realized the fate of *Columbia*, they began to implement the shuttle contingency plan, establishing an external accident investigation committee immediately and sharing information publicly as soon as it became available. NASA convened an initial press conference within a few hours of the accident, and senior shuttle and other agency officials conducted daily press updates for two weeks. NASA invited members of the news media to see accident investigators' progress in laying out salvaged wreckage on a full-sized outline of the shuttle in a hangar at Kennedy Space Center.[5] Shuttle officials even invited suggestions from the public and NASA employees for improving the vehicles' safety. Over a year's time, the agency received nearly 2,700 ideas on topics ranging from inspecting the orbiters in flight to rescuing endangered crews.[6]

As happened after *Challenger*, polling numbers showed an uptick in the percentage of Americans who believed that the space program warranted a funding increase and a drop in the number of those who thought it deserved less or no funding.[7] Adults and children sent letters and emails to NASA encouraging the shuttle program and human space flight to continue. Milt Heflin, Johnson Space Center flight director chief when the *Columbia* accident occurred, recalled a poignant letter from a man who wished NASA well and asked the agency to "please take us with you" as it prepared to return the shuttle to flight.[8] Applications for the Educator Astronaut opportunity kept rolling in.

The latest accident also resurfaced questions in the media about the future of human space flight. This time, however, NASA's proactive communications about the *Columbia* accident and investigation saved

FIGURE 8.1. Cartoons such as this one by Rob Rogers expressed a resolve to move forward with US human space exploration following the loss of the space shuttle *Columbia* and her crew. © 2003 by Rob Rogers, reproduced by permission from the artist.

the agency from the scathing reactions from journalists it had endured following the *Challenger* incident. Rather than levying harsh criticism at NASA, the media pondered the wisdom and purpose of flying humans aboard the shuttle. Some editorial cartoons saluted the fallen astronauts or recognized that the accident had not broken the American spirit of space exploration (see fig. 8.1). In an effort to suggest that it was time for NASA to change course in space exploration, other cartoons and newspaper articles cast the shuttle as an aged and overextended vehicle saddled to Earth orbit while nimble robotic explorers surveyed the Martian surface.[9] The *New York Times* advocated that NASA redirect its resources to projects aimed at "the sheer thrill of exploration and new discoveries."[10]

Seven months of examining the circumstances around the accident led the Columbia Accident Investigation Board (CAIB) to reach conclusions concerning its causes and to offer recommendations for moving forward with NASA's human space flight program. Made up of representatives from NASA, the navy, the air force, universities, industry, and aviation safety organizations, the CAIB suggested that the accident's causes were rooted deep in the shuttle's history and tied back to NASA's efforts to make the vehicle broadly relevant. They pointed out that a lack of high-level prioritization of human space flight since Kennedy's 1961 direction to send astronauts to the moon had forced NASA to "participate in the give and take of the normal political process" and to "gain the support of diverse constituencies" to obtain resources for the space shuttle and space station programs. Moreover, the CAIB noted that "the increased complexity of a Shuttle designed to be all things to all people" had contributed to its risks and that NASA had mischaracterized the shuttle as operational when it remained an experimental vehicle.[11] The CAIB strongly supported the shuttle returning to flight provided that NASA made near-term technical and managerial changes to reduce the chances for further accidents and recertified the shuttle to operate safely beyond 2010.

NASA and the Bush administration recognized that the nation stood at another critical juncture with respect to the future of the shuttle and NASA's human space flight program. Notwithstanding the CAIB's critique, the shuttle had vastly broadened participation and the utility of human space flight over three decades, in service to the nation and its citizens. It was also crucial to the completion of the International Space Station. But as the CAIB pointed out, it had proven to be a far more complex system with greater risks than NASA had ever anticipated. NASA's top leadership discussed with Bush administration officials whether continuing to send astronauts on missions into orbit around Earth was worth the risk to people's lives. Human space flight enthusiasts at NASA had long envisioned activity in low Earth orbit as part of a progression to send people to explore worlds beyond; agency and administration personnel felt that refocusing human missions on this ambitious aim would better justify the hazard.

President Bush's Office of Management and Budget maintained that any new human space flight pursuit would need to be undertaken without large budget increases, lest the plan wither in Congress like the tremendously costly Space Exploration Initiative proposed by Bush's father, President George H. W. Bush. The shuttle consumed some $2 billion of NASA's $16 billion annual budget. The administration decided that the best course of action was to terminate the shuttle program and create a new vision for space exploration. On January 14, 2004, Bush made a speech at NASA Headquarters to announce that the space agency would pursue a new plan for human space flight. Humans, together with robotic probes, would move beyond Earth orbit to explore the moon, Mars, and the solar system beyond. Completing the International Space Station and focusing US research aboard it on understanding the impacts of the space environment on human health would remain a critical stepping-stone toward realizing this vision. NASA would focus the shuttle on finishing station assembly and retire the orbiters thereafter, freeing resources needed to fund the new exploration program.[12]

As NASA commenced planning activities to set the new human space flight policy in motion, the July 2005 launch of STS-114 returned the shuttle to service and kicked off the final era in the vehicle's long history. Now focused exclusively on completing International Space Station construction, the shuttle transitioned to becoming the support vehicle that NASA human space flight proponents had envisioned while the Apollo missions were still flying. The agency no longer depended on securing the active involvement of various segments of the public to ensure the shuttle's viability. NASA flew only career astronauts from that point forward, albeit crews of women and men that were more racially, ethnically, and professionally diverse than ever before. Meanwhile, shuttle cargo space was almost entirely devoted to hardware required to build and supply the space station. NASA's new administrator, Michael Griffin, terminated the GAS and Hitchhiker payload programs shortly after his arrival in 2005, citing the lack of available room.

As space station assembly proceeded and NASA studied options for human exploration of the solar system, budget limitations kept the agency from having another launcher ready until a few years after the shuttle

FIGURE 8.2. Visitors to Kennedy Space Center watch as space shuttle Atlantis launches to the International Space Station with a six-member crew on STS-132 in May 2010. Photograph from NASA.

program was projected to end. Consequently, the shuttle would be the primary visible element of NASA's human space flight endeavors throughout its remaining life. Recognizing the shuttle's longstanding popularity, the agency strived to connect enthusiasts, the education community, and others to the vehicle in ways that continued to convey that it belonged to all Americans.

NASA's public engagement efforts took multiple forms. Although NASA stopped issuing car passes to view launches after the September 11, 2001, terrorist attacks, officials continued to welcome members of the public on organized bus tours to see the shuttle lift off (see fig. 8.2). The agency issued an open call for ideas for potential activities and experiments that educator-astronauts could complete in orbit and attracted tens of thousands of recommendations within just a few months.[13] NASA also used all available communications media to connect with the public and became an early adopter of social media. The agency's "tweeting" astronauts provided a new form of public access to the experiences of space flyers on shuttle missions, and they emerged as some of the most followed accounts on social media platforms. Beginning in 2009, the agency invited Twitter users and bloggers to the first in a series of "Tweet-ups" (later called "NASA Socials") that allowed them to attend shuttle launches, tour NASA facilities, and meet NASA personnel. Officials hoped that by giving active social media users access to NASA, participants would share in the excitement of human space flight, and in turn, communicate their experiences to new audiences online.[14]

The agency also encouraged broad public participation in certain cultural aspects of shuttle and station missions. In 2009 NASA held an online poll to name a space station node. The public response was tremendous, largely because late night television host Stephen Colbert had urged viewers to pass over NASA's four suggested names and put forward his surname as a write-in vote. ("Colbert" earned the highest number of votes, but NASA went with "Serenity," the most popular of the names it had suggested. The agency, however, responded in good fun by naming an exercise machine aboard the station the Combined Operational Load Bearing External Resistance Treadmill [COLBERT]). NASA also ran contests in 2010 to select the music to wake up the shuttle crew on STS-133 and to compose original wakeup songs for the STS-134 astronauts. The former contest garnered 2.5 million votes, and the latter received 1,350 submissions from sixty-three countries.[15]

Although the vehicle's purpose and accessibility had shifted over time, NASA officials wanted to honor and commemorate the shuttle as the people's spaceship through the program's close, preserving the imaginary that had been linked with the vehicle for so long. A few years after President Bush called for the shuttle's termination, NASA began to contemplate how to dispose of the hardware after the program ended. The agency developed a process for educational institutions to apply to acquire major elements such as flight trainers. It promoted the availability of smaller items such as leftover thermal protection tiles and packages of shelf-stable crew food to schools and museums.[16]

What to do with the orbiters, however, warranted further consideration. Administrator Griffin initially suggested that they should be displayed at NASA centers connected closely with the shuttle program. But the costs that NASA would have to bear to prepare and maintain the displays prompted the agency to take a different approach. Instead, it solicited proposals from museums and educational institutions around the country that would be willing to cover the costs to take in an orbiter. NASA would transfer one orbiter, *Discovery*, to the Smithsonian Institution, which by long-term agreement with the agency had the first rights to any NASA artifact. Twenty-one institutions vied for the remaining spacecraft. Charles Bolden, who became NASA's next administrator in 2009, prioritized those that had the highest annual attendances, regional populations, and accessibility by international visitors. On April 12, 2011, the thirtieth anniversary of the shuttle's first flight, Bolden announced the three additional institutions that would receive orbiters—Kennedy Space Center, *Atlantis*; California Science Center, *Endeavour*; and the Intrepid Sea, Air, and Space Museum in New York, *Enterprise*. These institutions, Bolden stated, would "provide the greatest number of people with the best opportunity to share in the history and accomplishments" of the shuttle program.[17]

The agency worked with its partners and authorities in those localities to give the vehicles grand welcomes for all to witness in 2012 and 2013. Millions watched as *Discovery* majestically circled the Washington, DC, area atop the 747 carrier aircraft before landing at Dulles International Airport for transfer to the Smithsonian National Air and Space Museum (see fig. 8.3); as *Enterprise* sailed serenely up the Hudson River by barge to a new home in New York City (see fig. 8.4); as *Endeavour* paraded through

▶ FIGURE 8.3 (TOP). Mounted on a 747 Shuttle Carrier Aircraft, space shuttle *Discovery* flies over Washington, DC, on April 17, 2012, before landing at Dulles International Airport for retirement at the Smithsonian National Air and Space Museum Steven F. Udvar-Hazy Center in Chantilly, Virginia. Photograph from NASA.

▶ FIGURE 8.4 (BOTTOM). Space shuttle Enterprise rides on a barge past the Statue of Liberty on its way to the Intrepid Sea, Air and Space Museum in New York City on June 6, 2012. Photograph from NASA.

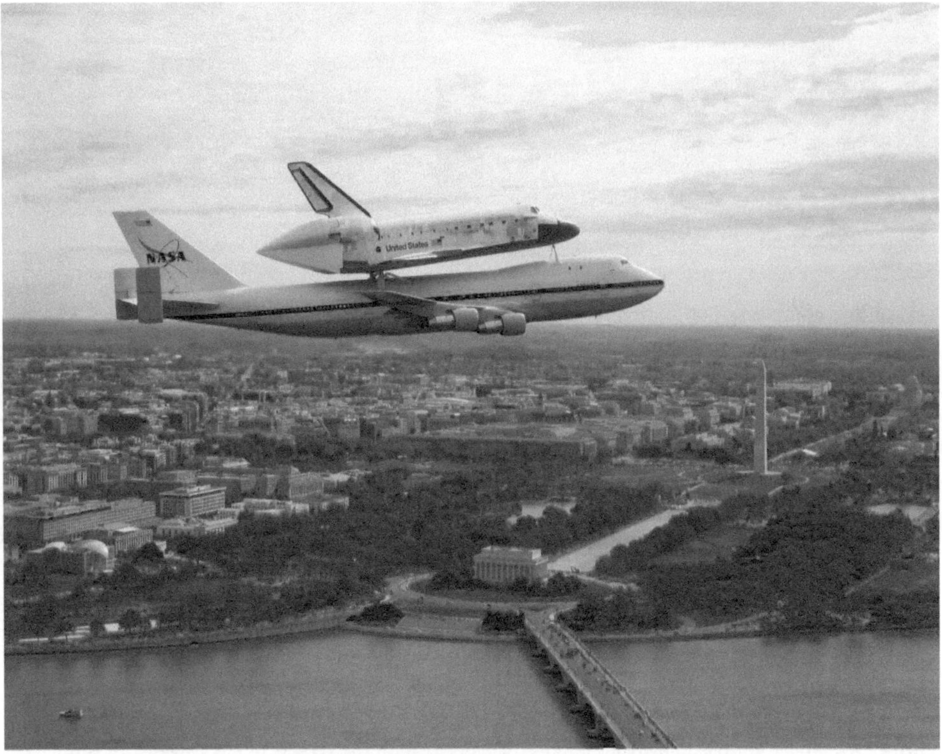
NASA
United States

NASA
United States
Enterprise
SHUTTLE ENTERPRISE
TO INTREPID
WEEKS 297
NYPD

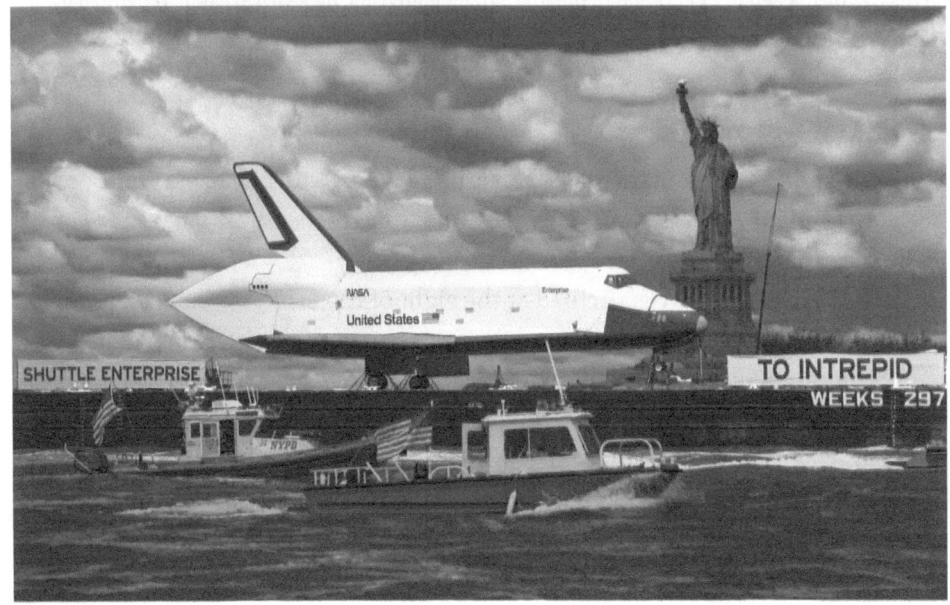

FIGURE 8.5. Space shuttle *Endeavour* makes its way through the streets of Los Angeles en route to the California Science Center on October 13, 2012. Photograph by Pedro Szekely; reprinted by permission from the photographer.

the streets of Los Angeles to reach the California Science Center (see fig. 8.5); and as *Atlantis* traversed ten miles across Kennedy Space Center to the center's visitor complex, where she was greeted with the fanfare of fireworks (see fig. 8.6).

While the crowds thrilled at the sight of the orbiters, the American citizenry, as always, remained far from unanimous about the value of the shuttle and NASA's human space flight pursuits. Opinion polls taken as shuttle flights came to an end revealed that more than one-third of Americans still did not think the shuttle had been a prudent use of tax dollars.[18] Others lamented that the vehicle had not opened space to the masses to the extent that NASA had projected in the 1970s. At the same time, the shuttle had galvanized participation from communities

FIGURE 8.6. Crowds and bursts of fireworks welcome space shuttle *Atlantis* to the Kennedy Space Center Visitor Complex on November 2, 2012. Photograph from NASA.

ranging from life scientists to educators to artists. Year after year over the shuttle's lifetime, through its successes and tragedies, more Americans approved of the shuttle as a national investment than ever approved of the Apollo program, even when the first astronauts reached the moon in 1969.[19] The number of Americans disappointed that the shuttle program was ending outweighed those who supported its termination by nearly three to one. Those who witnessed and attended the orbiter retirement festivities expressed feelings of nostalgia, with many equating the shuttle era's completion to the demise of the American space program.[20] Visitors to NASA facilities, concerned that the space agency would soon close its doors forever, bought up pins, patches, and other shuttle souvenirs in attempts to own a piece of the program's history.

As with Apollo, national pride and technological exuberance had much to do with the sentiments that Americans expressed at the end of the shuttle era. But their views were also shaped by NASA's early vision of a shuttle accessible to many and the agency's resulting commitment to offering numerous opportunities for the public to engage with the space transportation system. Even though these efforts were tempered as NASA balanced numerous competing interests over the vehicle's service life, they unequivocally played a significant role in defining the shuttle as a beloved American technological icon.

Conclusion

From a technical perspective, the space shuttle program began as an experimental effort to make sending humans and cargo to space routine and less costly through a partially reusable spacecraft. Such a vehicle, NASA officials reasoned, would extend capabilities that had been acquired through the Mercury, Gemini, and Apollo programs while making access to space more affordable than expendable launchers allowed. For four decades, NASA worked to meet the technical requirements the shuttle's designers had set for it. Agency officials publicly characterized the shuttle as an operational vehicle after flying four missions and in many respects convinced themselves and the American people that it had become a well-understood machine. The shuttle lofted dozens of satellites and space probes to Earth orbit, retrieved some for return, and repaired others. It carried small experiments as well as full laboratory modules and achieved the purpose NASA had originally envisioned for it: aiding in assembling a space station in Earth orbit. Yet, throughout its service life, the shuttle constantly tested the limits of its performance capabilities and, as two catastrophic accidents revealed, remained subject to risk and error.

The shuttle also entailed an experiment of another kind for NASA. It marked the agency's trial of new philosophies and forms of interaction with the American public to demonstrate the value of human space flight. NASA had, from its start, seen showing a commitment to serving the nation's people and seeking their support as instrumental to persuading Congress to invest in its activities. Operating as though it attended to one large attestive public, however, ultimately proved unfruitful for NASA during its quest for the moon. Consequently, although the shuttle's designers thought conservatively about its capabilities, NASA's leadership and public affairs officials were eager to make the shuttle economically and socially—and in turn, politically—viable and thus presented it as a versatile vehicle that promised something meaningful for everyone. The success of human space flight, they surmised, required more than solving technical challenges and publicly sharing the results of those efforts; NASA had to become versed in customer relations and meet the needs of many.

To a considerable extent, NASA's experiment yielded the results the agency sought. The promise that the shuttle would broaden access to new users of space and welcome new types of flyers appealed to President Richard Nixon and triggered an outpouring of interest and petitions from students, artists, and many others who wanted to be part of the new human space flight program. Personnel in NASA's space flight, education, and public affairs offices scrambled to create initiatives that would push boundaries, constituting new forms of public participation for the agency and democratizing access to space commensurate with the vehicle's known performance capabilities. Subsequent presidents and influential members of Congress championed several of these initiatives. For advocates, the shuttle became not just a spacecraft to be displayed and discussed but a people's spaceship, where the objectives of human space flight aficionados within NASA and the interests of individuals from all walks of life met. NASA's efforts to realize this vision resulted in the shuttle becoming more popular among Americans than the Apollo program ever was.

Even so, expanding the ranks of participants in the shuttle program to the extent that some inside and outside of the agency had envisioned proved challenging, and the experiment indicated limits to NASA's ability—and willingness—to simultaneously support multiple publics' interests in the shuttle. The agency was never able to achieve the flight rate it had promised in the shuttle's development stage. Given the vehicle's restricted availability, some NASA officials raised concerns about how to reconcile broadening access to it with what priority should be given to those with longstanding connections to the space program, including career astronauts and professional scientists, who favored preserving or creating opportunities for themselves. Under pressure from the Reagan administration and again in the late 1990s, agency officials struggled with how to avoid competition between the shuttle and companies interested in building space businesses. Meanwhile, the 1986 *Challenger* disaster damaged the agency's credibility and esteemed image with some in the media and Congress. Whereas NASA had seen opening access to the shuttle as the solution to garnering the political support it needed to sustain the program, the agency reacted by doing the opposite and constricting efforts to involve more segments of society as shuttle users and flyers. After NASA secured Congress's approval of the International Space Station and tasked the shuttle with its construction, and especially after the shuttle program was slated for termination following the *Columbia* accident, increasing direct public participation did not matter as much to the agency because it no longer needed to rationalize the vehicle.

In a democracy, federal technoscientific agencies have tough jobs, tasked with fulfilling lofty national goals and making decisions about how to manage research and technology development and use in the face of limited budgets and potential risks. From its inception and compounded by its early colossal technological success with Apollo, NASA has been held to extraordinarily high standards and demands, most notably by political stakeholders, the media, and those whose livelihoods benefit most from its expenditures, including the space science community and the aerospace industry. Collectively, they have propagated expectations for the agency to undertake novel and complex space projects, maximize discovery and benefits of space flight, create excitement and adventure, and perform flawlessly in the face of the unforgiving space environment, all while justifying spending many billions of taxpayer dollars annually.

In the case of the shuttle, the next major post-Apollo human space flight program, NASA felt tremendous pressure to live up to these ideals. From the program's outset NASA officials embraced the strategy of engaging more publics in a variety of ways, thereby demonstrating service to the American people while striving to build constituencies for the vehicle. Indeed, agency officials correlated widespread public support with political support, recognizing the latter's paramountcy for continuing humanity's journey into the solar system that so many of them dreamed of.

Charting NASA's experience with public engagement over the shuttle's lifetime, however, illuminates that the agency's quests for public and political support were not entirely synergic in practice. Officials in the White House or Congress sometimes reacted to the concerns of established or powerful industry players above heeding interests equitably and opening the shuttle to all. When they favored some groups, or when the agency's reputation or long-range human space flight plans seemed to be on the line, NASA often capitulated to those politicians' and actors' preferences, even if it meant retrenching from expanding public access to the shuttle as fully as it might have. NASA nonetheless never stopped showcasing the shuttle, continuing to believe that public engagement was, at least to some degree, foundational for the human space flight program's perpetuation.

The shuttle orbiters and associated hardware took their places as museum pieces around the country in 2012 (see fig. C.1). Their popularity endures, as they attract and inspire millions of people annually who view and learn the story of the remarkable spaceships and what NASA, in collaboration with so many Americans and international partners, achieved over the program's forty-year history. Each orbiter, in silent repose, holds insights about the opportunities and benefits, as well as the challenges and limitations, associated with designing space programs that engage various publics. As museum visitors behold the majestic spacecraft, NASA is busily working to return astronauts to the moon via the Artemis program in preparation for future human missions to Mars. In these final pages, I offer some personal thoughts on how the shuttle's lessons about broadening access, democratizing space flight, and otherwise creating value for Americans could guide NASA in planning for public engagement

FIGURE C.1. Space shuttle *Atlantis* on display at the Kennedy Space Center Visitor Complex as if in flight, rotated forty-three degrees with payload doors open. Photograph from NASA.

with lunar and Mars exploration that is purposeful for the agency and participants alike.

Above all, examining NASA's public engagement with the shuttle reveals that establishing a vision for, messaging about, and opportunities associated with human space flight to meet different groups' needs and interests is key to widespread public acceptance of its initiatives. But the shuttle experience also points to a need for NASA officials to reconsider and reframe the reasons the agency engages the public. In an environment in which entrenched stakeholders and national imperatives hold sway over elected officials' decisions, it is a fallacy to assume that political support for a space initiative will be heavily influenced by public opinion. This was the case during the shuttle era, and it is again proving true for Artemis, where successive administrations and Congresses have provided funds for the lunar program's development in response to powerful aerospace companies' interests and growing concerns of a national security threat from China as the Asian country reveals its space capabilities and plans.

NASA should continue to engage Americans—and global publics—in its human space flight efforts. However, the shuttle experience suggests that rather than equating public support with garnering appreciation for NASA's achievements to in turn secure federal resources to accomplish more, the agency would do better to reimagine the concept of "public support" toward more realistic and productive ends. Such a revised definition ought to encompass the extent to which NASA, through its various engagement approaches, connects its ambitions meaningfully and purposefully with its many publics. Two principles NASA could focus on to achieve this include how it serves its publics and how it involves them substantively in its human space flight programs. As the shuttle experience showed, adopting these alternative means for valuing publics can lead to effective methods to engage them while enriching NASA's programs in a variety of ways.

Serving publics well requires gaining their buy-in and meeting their needs. The shuttle's sociotechnical imaginary of an accessible vehicle that would satisfy many people was highly effective in generating public approval of and participation in the program. Many Americans got excited about a new, groundbreaking project in which they could see themselves or their interests being served. NASA has produced a strategic framework of goals and objectives for sustained human exploration of the moon and Mars. Drawing on input solicited from industry, academia, and international partners, the agency cites reasons for this pursuit that include scientific discovery, economic benefits, and American leadership in space, and inspiration of a new generation of explorers.[1] According to a 2021 poll, however, only a third of Americans believe returning astronauts to the moon should be a priority for NASA.[2] This statistic indicates that most do not see the program's value. Honing the rationales to articulate whom Artemis will benefit and how in practice as well as refining how they are communicated will be important to enabling various publics to recognize the program as a worthwhile investment that improves their lives and the world around them.

In a similar vein, public service also entails considering how the new human space flight program will comport with societal and cultural values. NASA's movement toward increasing the diversity of the astronaut corps that would fly aboard the shuttle was a response to societal changes and pressures. NASA has proclaimed that the first Artemis mission to land on the moon will introduce new demographics to the lunar surface: a woman and a person of color. Following the European Space Agency's announcement of plans to select an astronaut with a physical disability, NASA sponsored a study in 2021 to explore the feasibility of sending people with physical limitations into space.[3] In 2023 NASA held a workshop with philosophers, sociologists, and other social scientists to identify societal, ethical, and cultural issues to consider in going to the moon and beyond.[4] For example, to what extent will NASA reconcile lunar exploration plans with the moon's sacred status among many Indigenous peoples? And how will NASA decide what artifacts of exploration are worthy of preserving as heritage sites on the lunar surface? The agency will need to establish policies and processes, many in collaboration with international space agencies, to address such concerns.

Public service also entails engaging multiple publics with NASA's programs in ways they appreciate. The agency worked hard from the shuttle program's start to create approaches that met various groups' interests. Many people enjoyed opportunities to witness a launch or thrilled at the ability to listen to mission control's communications with the astronauts. Others were eager to fly research projects into space. NASA could conduct surveys through its website, social media, or network of informal education institutions to gauge public satisfaction with how the agency serves their interests and engages them in its moon and Mars exploration programs. Officials could review the feedback and implement appropriate changes.

One thing that is clear is that today, perhaps even more so than in the shuttle era, many people can and want to participate directly in space activities. Both the Apollo and shuttle programs spawned a tremendously capable aerospace industrial base in the United States that is developing the Orion crew vehicle, the Space Launch System booster, and other hardware components for the Artemis program. The earlier human space flight initiatives also provided inspiration for companies like SpaceX and Blue Origin to emerge more recently. The latter companies have secured contracts for systems that will take astronauts from lunar orbit to the moon's surface and, along with several other firms, are building spacecraft to deliver science and technology payloads to the moon. Small businesses, university teams, K-12 students, and others are creating experiments to support lunar exploration using various sources of NASA funding. All the while, private entities are now demonstrating the ability to launch cargo and human crews, including paying customers, into suborbital and orbital space. These changes have prompted NASA to move into the role of a customer of services supporting the International Space Station and to set policies to encourage commercial uses of the station and build an economy in low Earth orbit.

Still, one doesn't need to stake a career in aerospace to contribute in valuable ways to space exploration. This was true in the shuttle era, but even more so today as NASA finds itself in a world in which information technology networks connect and empower more people to yield social and technoscientific impact. This development, accelerated by the COVID-19 pandemic's influence in shifting attitudes and habits regarding work, has unleashed a "gig economy" of freelancing individuals availing their skills and expertise to organizations or disciplines that might use them. Climate change and the environmental justice movement have coaxed the civic-minded, students, teachers, and others to get involved in projects to monitor air and water quality, pollinator and mosquito habitats, and more in their communities. Indeed, how technological innovation and science happen is evolving rapidly and expanding to include legions of people previously uninvolved in these pursuits.

NASA has already made a start at recognizing these changes as opportunities to harness the ingenuity and abilities of American citizens and people around the globe. Like other US government agencies and technoscientific institutions, the agency has created public prize competitions, challenges, and citizen science projects to enhance its space research and technology development pursuits. This type of public participation extends NASA's workforce, allowing the agency to tap into ideas from different disciplines that offer fresh starting points, and sometimes more complete solutions, for problems, including those related to human space flight. NASA has made awards to computer coders for designing applications to improve tracking of astronauts' food intake and to an independent inventor for improvements to the flexibility of spacesuit gloves. The agency has invited public competitors to demonstrate systems for storing and distributing power on the moon and has held a contest with a track for K–12 students to suggest concepts for toilets that could operate on the lunar surface. NASA's use of these approaches has grown over the years but is still small compared to its reliance on grants and contracts for drawing on external innovation, so the agency could scale its use of these opportunities to further public engagement and meet Artemis needs.

Likewise, the agency could involve more people in human space flight planning. So far, decisions on goals and objectives for Artemis and human exploration of Mars have been driven largely by NASA and its international and industry partners. NASA could engage the broader public in evaluating the meaningfulness of the rationales the agency has proposed for the Artemis program and invite public comment during reviews of its moon-to-Mars architecture. To provide rich grounds for feedback, exchange of ideas, and understanding of social and cultural concerns, the agency could host in-person or online forums where members of the public could discuss and respectfully debate the opportunities and challenges Artemis presents. NASA forged a successful model for this activity when it commissioned a consortium of academic and informal education institutions to hold meetings to solicit public views on the agency's plans to search for potentially hazardous asteroids and to involve astronauts in exploring an asteroid.[5]

As was the case with the shuttle, public engagement possibilities abound, and there are many options to substantively involve people with little technical background. Americans—or perhaps global citizens—could be invited to vote on landing sites from among the options identified by NASA's science and engineering teams. NASA could hold a public competition to design Artemis crew mission patches. And just as NASA reconsidered who ought to make up its astronaut corps in the shuttle era, citizens could advise the agency on what qualifications and characteristics its new space travelers should possess. These and still more opportunities for involvement with lunar and Mars exploration could make public participants partners with the agency and offer them a stake in NASA's programs and ensure that those programs reflect a breadth of values.

Public engagement, done effectively, requires a great deal of energy, creativity, political savvy, and patience. It requires a willingness to understand and accept the validity of views of people of many walks of life—even when they do not match one's own—and to keep at the forefront consideration of various interests and a focus on service. Revising current processes and considerations in Artemis and Mars exploration program development to incorporate more players would no doubt add complexity to an already complicated decision and operating space.[6] It could even feel to NASA officials and longstanding space program developers like an existential threat: what would inviting broader public involvement say about their abilities to do their jobs and make reasoned judgments? How would they cope with preferences that ran afoul of their own senses of what is right to do for human space flight? Indeed, public engagement in the most genuine sense entails giving up some control and trusting that listening to and collaborating with others can enhance program outcomes.

At NASA, the importance of making space for everyone will only grow. As private companies pursue increasingly ambitious feats, opening space tourism (granted, for now, only to the few who can afford the exorbitant price tags) and even seeking to reach the moon independently, the agency will face questions about the value of the federal mission if Artemis endures snags and delays. But government agencies continue to have roles to play in investing in exploring new scientific frontiers that companies, focused on nearer-term returns, do not dare to pursue. NASA is fulfilling that role in its quest for the moon and Mars. Its experience engaging multiple publics with the space shuttle with an eye toward meeting their many interests should encourage the agency to define Artemis's success in part in terms of how effectively it connects with Americans as well as global citizens. At the very least, greater public participation will ensure that the United States' future in space will be a rich and democratic one, evoking the ideal of a people's space program that was embodied in the shuttle several decades ago and offering a more meaningful future in the cosmos for all.

Notes

Preface

1. John M. Logsdon, ed., *Exploring the Unknown: Selected Documents in the History of the US Civil Space Program*, NASA SP-4407, 8 vols. (Washington, DC: US Government Printing Office, 1995–2008).

Introduction

1. NASA, "NASA Transfers Space Shuttle to NASM," video of ceremony transferring *Discovery* to National Air and Space Museum, Chantilly, VA, April 19, 2012, posted April 20, 2012, https://www.youtube.com/watch?v=QSULwrdHF_g. Bolden's remarks begin at 36:05.

2. NASA, "NASA Transfers Space Shuttle to NASM." Dailey's remarks begin at 24:35.

3. It is difficult to find the proper vernacular to use when writing about NASA's engagement of those outside of government and industry space program and policy developers. "The public," "the general public," and "the American public" are terms that government organizations, politicians, pollsters, media outlets, marketing experts, and others have embraced to connote an undefined, indistinct grouping of people. NASA has also often used these blanket terms. However, these terms imply a homogeneity that does not exist in the United States, let alone throughout the entire world. In this book, I prefer to use "publics" to acknowledge the diversity of the people NASA reached, even if they were not uniquely identified by the agency. I do, however, rely on "the public" (and variants, including "public" as an adjective) to echo NASA's or others' use of these terms when I wish to highlight their thinking or where references to people without specificity seem appropriate. I also use terms such as "Americans" or "citizens" when referring to people within the United States without specificity. Where they are known, I name the specific people or groups with which NASA engaged—such as business owners, teachers, or *Star Trek* fans. Lastly, I use the term "public engagement" to refer generally to NASA's reach to those beyond traditional space program developers.

4. See, for example, James L. Kauffman, *Selling Outer Space: Kennedy, the Media, and Funding for Project Apollo, 1961–1963* (Tuscaloosa: University of Alabama Press, 1994); Mark E. Byrnes, *Politics and Space: Image Making by NASA* (Westport, CT: Praeger, 1994); Kristen A. Starr, "NASA's Hidden Power: NACA/NASA Public Relations and the Cold War, 1945–1967" (PhD diss., Auburn University, 2008); Bruce V. Lewenstein, "NASA and

the Public Understanding of Space Science," *Journal of the British Interplanetary Society* 46 (1993): 251–54; Linda Billings, "Fifty Years of NASA and the Public: What NASA? What Publics?" in *NASA's First 50 Years: Historical Perspectives*, ed. Steven J. Dick, NASA SP-2010-4704 (Washington, DC: US Government Printing Office, 2010), 151–81; and David Meerman Scott and Richard Jurek, *Marketing the Moon: The Selling of the Apollo Lunar Program* (Cambridge: MIT Press, 2014). Roles for others outside this sphere are discussed in Howard E. McCurdy, *Space and the American Imagination* (Washington, DC: Smithsonian Institution Press, 1997); Michael A. G. Michaud, *Reaching for the High Frontier: The American Pro-space Movement, 1972–84* (New York: Praeger, 1986).

5. Richard C. Levy, "Everybody Is Lining Up for Space," *Parade*, December 30, 1979, 4–5.

6. Ann P. Bradley, "NASA's New Space Flight Participant Program," August 27, 1984, folder 19758, NASA Historical Reference Collection, NASA Headquarters, Washington, DC.

7. Hans Mark, *The Space Station, A Personal Journey* (Durham, NC: Duke University Press, 1987), 57–58.

8. NASA, *Wings in Orbit: Scientific and Engineering Legacies of the Space Shuttle*, ed. Wayne Hale, Helen Lane, Gail Chapline, and Kamlesh Lula, NASA SP-2010-3409 (Washington, DC: US Government Printing Office, 2010), 9. Emphasis added.

9. In using terms such as "American citizenry," "American citizens," and "Americans" as I explore the relationship of the US government to the people it serves, I do not discount that NASA's outreach activities likely also had an impact on US residents without legal citizenship status.

10. Just as the American public is not a homogeneous entity, NASA is not a monolithic institution. The agency comprises tens of thousands of US government civil servants and contractors spread over ten NASA field centers throughout the country. Individuals across a range of NASA offices—shuttle management, public affairs, education, science, and safety and mission assurance, to name the key ones—engaged with the nation's people. When talking about "NASA," I aim to be explicit where possible about who participated, thought, and acted in what way.

11. Vannevar Bush, *Science, The Endless Frontier: A Report to the President* (Washington, DC: Government Printing Office, 1945).

12. Sheila Jasanoff and Sang-Hyun Kim, "Containing the Atom: Sociotechnical Imaginaries and Nuclear Power in the United States and South Korea," *Minerva* 47, no. 2 (2009): 119–46.

13. Yaron Ezrahi, *The Descent of Icarus: Science and the Transformation of Contemporary Democracy* (Cambridge, MA: Harvard University Press, 1990). Ezrahi explains how NASA created an attestive public around the Saturn V rocket (Ezrahi, 42). When visitors to Kennedy Space Center were taken to see it and heard from a NASA tour guide that it was built by companies in states from which they hailed, there was a "thunder of applause" from "proud citizens" who felt that they shared "a piece of the action."

Through "the language and rhetoric of democratic participation," Ezrahi states, the Saturn V "emerges suddenly as a monument to the ingenuity of an entire people, the creativity and contributions of many private American citizens and firms spread all over the United States."

14. National Aeronautics and Space Act of 1958, Pub. L. No. 85-568 72 Stat. 426-2 (1958).

15. See, for example, Don K. Price, *The Scientific Estate* (Cambridge, MA: Harvard University Press, 1967); Dorothy Nelkin, "The Political Impact of Technical Expertise," *Social Studies of Science* 5, no. 1 (1975): 35–54; and Martin Lengwiler, "Participatory Approaches in Science and Technology: Historical Origins and Current Practices in Critical Perspective," *Science, Technology, and Human Values* 33, no. 2 (2008): 186–200.

16. Roger D. Launius, "Public Opinion Polls and Perceptions of US Human Space Flight," *Space Policy* 19, no. 3 (2003): 163–75.

17. On the design and development of democratic technologies, see Langdon Winner, "Do Artifacts Have Politics?" *Daedalus* 109, no. 1 (1980): 121–36; and Richard E. Sclove, *Democracy and Technology* (New York: Guilford, 1995). I consider the space shuttle a democratic technology in that although NASA wished to develop it, the agency recognized a need for it to appeal to various social groups to succeed.

18. In this book I focus on how a democratic, utilitarian imaginary shaped NASA's public engagement approaches with the shuttle from its outset and how the imaginary's impact shifted over time. In comparison, Valerie Neal's *Spaceflight in the Shuttle Era and Beyond: Redefining Humanity's Purpose in Space* (New Haven, CT: Yale University Press with Smithsonian National Air and Space Museum, 2017) identifies multiple framings for human space flight espoused during the shuttle era by NASA, policy makers, the media, and others. Several of the themes Neal raises come up peripherally in my book.

19. John M. Logsdon, "The Space Shuttle Program: A Policy Failure?" *Science* 232, no. 4754 (1986): 1099–105; John M. Logsdon, *After Apollo? Richard Nixon and the American Space Program* (New York: Palgrave Macmillan, 2015), 295–99.

20. Traci Watson, "NASA Administrator Says Space Shuttle Was a Mistake," *USA Today*, September 28, 2005, 1A.

Chapter 1: Building a National Vision and Public for Space Flight

1. Roger D. Launius, "Prelude to the Space Age," in *Organizing for Exploration*, vol. 1 of *Exploring the Unknown: Selected Documents in the History of the US Civil Space Program*, ed. John M. Logsdon, NASA SP-4407 (Washington, DC: US Government Printing Office, 1995), 20.

2. Wernher von Braun, "Crossing the Last Frontier," *Collier's*, March 22, 1952, 25.

3. H. J. E. Reid, Director, NACA, to NACA, "Research on Spaceflight and Associated Problems," August 5, 1952, folder 18674, NASA Historical Reference Collection, NASA Headquarters, Washington, DC (hereafter cited as NHRC/HQ).

4. National Aeronautics and Space Act of 1958, Pub. L. No. 85-568 72 Stat. 426-2 (1958).

5. NASA, Office of Program Planning and Evaluation, "The Long-Range Plan of the National Aeronautics and Space Administration," December 16, 1959, in Logsdon, *Organizing for Exploration*, 403–7.

6. T. Keith Glennan, *The Birth of NASA: The Diary of T. Keith Glennan*, ed. J. D. Hunley, NASA SP-4105 (Washington, DC: US Government Printing Office, 1993), xxiv.

7. NASA, Office of Program Planning and Evaluation, "Long-Range Plan."

8. Lyndon B. Johnson to the president, memorandum, "Evaluation of Space Program," April 28, 1961, in Logsdon, *Organizing for Exploration*, 427–29.

9. James E. Webb and Robert McNamara to the vice president, May 8, 1961, with attachment, "Recommendations for Our National Space Program: Changes, Policies, Goals," 13, in Logsdon, *Organizing for Exploration*, 446.

10. Webb and McNamara to the vice president, May 8, 1961, 8, in Logsdon, *Organizing for Exploration*, 444.

11. John F. Kennedy, "Special Message to the Congress on Urgent National Needs," May 25, 1961, John F. Kennedy Presidential Library and Museum, http://www.jfklibrary.org/Research/Research-Aids/JFK-Speeches/United-States-Congress-Special-Message_19610525.aspx.

12. National Research Council, Space Science Board, *A Review of Space Research* (Washington, DC: National Academy Press, 1962).

13. Philip H. Abelson, "Manned Lunar Landing," Science 140, no. 3564 (1963): 267.

14. William D. Compton, *Where No Man Has Gone Before: A History of Apollo Lunar Exploration Missions*, NASA SP-4214 (Washington, DC: Government Printing Office, 1989), Appendix 2: Apollo Funding History, http://www.hq.nasa.gov/pao/History/SP-4214/app2.html#1966.

15. Dwight D. Eisenhower, "Farewell Address," January 17, 1961, Dwight D. Eisenhower Presidential Library, https://www.eisenhowerlibrary.gov/research/online-documents/farewell-address.

16. Atomic Energy Act of 1946, Pub. L. No. 79-585, 60 Stat. 755 (1946), 12.

17. Dwight D. Eisenhower to the Congress of the United States, April 2, 1958, in *Hearings on H.R. 11881, Day 1, Before the Select Comm. on Astronautics and Space Exploration*, 85th Cong. 3–5 (1958).

18. National Aeronautics and Space Act of 1958.

19. Select Comm. on Astronautics and Space Exploration, H.R. 12575, Establishment of the National Space Program, 85th Cong. (1958), H.R. Rep. No. 1770, 8–9.

20. Launius, "Prelude to the Space Age," 16.

21. Donald N. Michael, "The Beginning of the Space Age and American Public Opinion," *Public Opinion Quarterly* 24, no. 4 (1960): 575; Michael, "Society and Space Exploration," *Astronautics*, February 1958, 88–89. Emphasis in original.

22. Select Comm. on Astronautics and Space Exploration, H.R. 12575, Establishment of the National Space Program, 85th Cong. (1958), H.R. Rep. No. 1770, 36.

23. Anti-Lobbying Act, 18 U.S.C. § 1813 (1919).

24. Roger D. Launius, "Public Opinion Polls and Perceptions of US Human Space Flight," *Space Policy* 19, no. 3 (2003): 167.

25. Amitai Etzioni, *The Moon-doggle: Domestic and International Implications of the Space Race* (Garden City, NY: Doubleday, 1964).

26. Independent Offices Appropriations, 1962, *Hearing Before the Senate Comm. on Appropriations*, 87th Cong. 662 (1961) (statement of Hugh Dryden, deputy administrator of NASA).

27. Brian Duff, interview by John Mauer, April 24, 1989, transcript, Glennan-Webb-Seamans Project, box 12, folder 13, National Air and Space Museum Archives, Smithsonian Institution, Washington, DC.

28. Charles A. Biggs, interview by Rebecca Wright, August 1, 2002, transcript, NASA Johnson Space Center Oral History Project, Houston, TX, https://historycollection.jsc.nasa.gov/JSCHistoryPortal/history/oral_histories/BiggsCA/BiggsCA_8-1-02.htm.

29. Biggs, interview, August 1, 2002.

30. Brian Duff, interview by John Mauer, May 24, 1989, transcript, Glennan-Webb-Seamans Project, box 12, folder 16, National Air and Space Museum Archives.

31. Duff, interview, April 24, 1989.

32. NASA, "Public Affairs Program Review Document," NASA Headquarters, April 19, 1966, 2, folder 18179, NHRC/HQ.

33. Thomas O. Paine, "Public Interest in the Space Program: Appendix 5," Statement of Dr. Thomas O. Paine, Administrator, NASA, before the Committee on Aeronautical and Space Sciences, United States Senate, April 6, 1970, folder 6715, NHRC/HQ. Emphasis in original.

34. Jack King, interview by Patrick Moore, June 20, 2002, transcript, Kennedy Space Center History Project, Kennedy Space Center, FL.

35. "NASA Pays to Find Out if It Is Doing Its Job," *New York Times*, July 12, 1963, 3; "Public Be Dimmed," *Aviation Week*, September 23, 1963, 25; James L. Kauffman, *Selling Outer Space: Kennedy, the Media, and Funding for Project Apollo, 1961–1963* (Tuscaloosa: University of Alabama Press, 1994), 18.

36. NASA, "Apollo Information," May 18, 1967, folder 18179, NHRC/HQ.

37. Hugh L. Dryden, "The Citizen's Stake in Space Exploration," news release 62-236D, November 8, 1962, 2, https://historydms.hq.nasa.gov/sites/default/files/DMS/e000039118.pdf.

38. Julian Scheer to Mrs. Edward Levine, December 18, 1967, folder 1904, NHRC/HQ.

39. Julian Scheer to Mrs. E. G. Hunter, September 10. 1965, folder 1903, NHRC/HQ.

40. Julian Scheer to Mr. Terrence O'Neill, September 18, 1964, folder 1903, NHRC/HQ.

41. Wernher von Braun, "The Citizen in the Space Age," Carey Memorial Lecture, Lexington, MA, November 13, 1959, 3, NHRC online, accessed May 4, 2023, https://historydms.hq.nasa.gov/sites/default/files/DMS/e000041197.pdf.

42. Dryden, "Citizen's Stake in Space Exploration," 7.

43. NASA, *Exploring Space . . . Project Mercury* (Washington, DC: Government Printing Office, 1961), back cover.

44. James E. Webb to John A. Powers, August 9, 1963, folder 1732, NHRC/HQ.

45. NASA, "Apollo 11 Slows Down Washington," attachment to O. B. Lloyd Jr., letter to the editor, March 30, 1970, folder 18179, NHRC/HQ.

46. Julian Scheer to distribution, memorandum, "Protocol Plans for Apollo Manned Missions," June 17, 1968; Julian Scheer to distribution, memorandum, "Invitations to View Launch of Apollo 5," December 18, 1967; both in folder 1904, NHRC/HQ.

47. James E. Webb and Hiden T. Cox, "Accelerated Space Exploration: An Imperative for Americans," remarks for delivery at the Annual Convention, American Association of School Administrators, Atlantic City, NJ, February 18, 1962, NASA news release 62-34, 17, folder 18179, NHRC/HQ.

48. NASA, *Space: The New Frontier* (Washington, DC: Government Printing Office, 1963), 2.

49. Roger D. Launius, "First Steps into Space: Projects Mercury and Gemini," in *Human Space Flight: Projects Mercury, Gemini, and Apollo*, vol. 7 of Logsdon, *Exploring the Unknown*, 18.

50. "Secrecy in Space," *Washington Post*, August 27, 1965.

51. Duff, interview, April 24, 1989; NASA, "Public Affairs Program Review Document," 3; Thomas O. Paine, "Public Interest in the Space Program: Appendix 5," 4; "Comment Cards," news release NT-369, TWA-NASA Tours press packet, July 11, 1969, NHRC/HQ.

52. John P. Donnelly to Fletcher, memorandum (draft), October 28, 1971, folder 7638, NHRC/HQ.

53. Louis Harris, "Space Programs Losing Support," *Washington Post*, July 31, 1967.

54. 113 Cong. Rec.

Chapter 2: Bringing Human Space Flight Closer to Earth

1. National Aeronautics and Space Act of 1958, Pub. L. No. 85-568 72 Stat. 426-2 (1958).

2. Wayne Hale, author interview, April 12, 2013.

3. Thomas O. Paine to the president, memorandum, "Problems and Opportunities in Manned Space Flight," February 26, 1969, in *Organizing for Exploration*, vol. 1 of *Exploring the Unknown: Selected Documents in the History of the US Civil Space Program*, ed. John M. Logsdon, NASA SP-4407 (Washington, DC: US Government Printing Office, 1995), 513–19; T. A. Heppenheimer, *The Space Shuttle Decision: NASA's Search for a Reusable Space Vehicle*, NASA SP-4221 (Washington, DC: US Government Printing Office, 1999), 148.

4. Richard Witkin, "Agnew Proposes a Mars Landing," *New York Times*, July 17, 1969, 1, 22.

5. Robert C. Seamans Jr. to Spiro T. Agnew, August 4, 1969, in Logsdon, *Organizing for Exploration*, 522.

6. Space Task Group, *The Post-Apollo Space Program: Directions for the Future*, September 1969, in Logsdon, *Organizing for Exploration*, 523–24.

7. John Ehrlichman, *Witness to Power: The Nixon Years* (New York: Simon and Schuster, 1982), 144–45.

8. Space Task Group, *Post-Apollo Space Program*, 523–40.

9. Thomas O. Paine to the president, September 19, 1969, quoted in John M. Logsdon, *After Apollo? Richard Nixon and the American Space Program* (New York: Palgrave Macmillan), 80.

10. "A Troubled NASA Begins Another Decade, Impasse for Air Defense," *Space/Aeronautics*, October 1968, 27.

11. Richard Nixon, "Proclamation 3919—National Day of Participation Honoring the Apollo 11 Mission," July 16, 1969, *The American Presidency Project*, ed. Gerhard Peters and John T. Woolley, https://www.presidency.ucsb.edu/node/239616.

12. Richard Nixon to the vice president, the secretary of defense, the acting administrator, NASA, and the science advisor, memorandum, February 13, 1969, in Logsdon, *Organizing for Exploration*, 513.

13. Remarks of Vice President Agnew, July 7, 1969, attached to memorandum, Russell Drew to Members and Observers of the Space Task Group, July 29, 1969, quoted in Logsdon, *After Apollo?*, 60.

14. American Institute for Aeronautics and Astronautics, "The Post-Apollo Space Program: An AIAA View," May 20, 1969, quoted in Logsdon, *After Apollo?*, 60.

15. Harry Hess to Lee DuBridge, June 23, 1969, quoted in Logsdon, *After Apollo?*, 60–61.

16. Louis Harris, "Americans Still Question Space Budget," *Washington Post*, August 25, 1969, A2.

17. "A Poll Finds Public Lukewarm on Mars," *New York Times*, August 7, 1969, 20. Agnew and Paine had publicly floated the idea of a human mission to Mars before the Space Task Group report's September 1969 release.

18. Walter Sullivan, "Next: On to Mars?" *New York Times*, July 27, 1969, E1.

19. Ralph E. Lapp, "Send Computers, Not Men, Into Deep Space," *New York Times*, February 2, 1969.

20. NASA, *Astronautics and Aeronautics, 1969: Chronology on Science, Technology, and Policy*, NASA SP-4014 (Washington, DC: US Government Printing Office, 1970), 201.

21. Future NASA Space Programs, *Hearing Before the Senate Comm. on Aeronautical and Space Sciences*, 91st Cong. 47 (1969) (statement of Senator Mark Hatfield).

22. William J. Normyle, "Manned Mission to Mars Opposed," *Aviation Week & Space Technology*, August 18, 1969, 16.

23. "Kennedy Asks Space Slowdown," *Washington Post*, May 20, 1969, A13.

24. Joseph Morgenstern, "What's It to Us?" *Newsweek*, July 7, 1969, 68.

25. Michael A. G. Michaud, *Reaching for the High Frontier: The American Pro-space Movement, 1972–84* (New York: Praeger, 1986).

26. See, for example, "People Needed in Picture," *Sentinel Star* (Orlando, FL), September 16, 1973, folder 6715, NASA Historical Reference Collection, NASA Headquarters, Washington, DC (hereafter cited as NHRC/HQ).

27. Kim McQuaid, "Selling the Space Age: NASA and Earth's Environment, 1958–1990," *Environment and History* 12, no. 2 (2006): 140.

28. Herbert E. Krugman, "Public Attitudes toward the Apollo Space Program, 1965–1975," *Journal of Communication* 27, no. 4 (1977): 93; John Noble Wilford, "A Spacefaring People: Keynote Address," in *A Spacefaring People: Perspectives on Early Spaceflight*, ed. Alex Roland, NASA SP-4405 (Washington, DC: US Government Printing Office, 1985), 71.

29. Peter M. Flanigan to the president, memorandum, December 6, 1969, in Logsdon, *Organizing for Exploration*, 546.

30. Thomas Paine, memorandum for the record, "Meeting with The President," January 22, 1970, 1, folder 12578, NHRC/HQ. The quoted phrases are Paine's summary of Nixon's comments.

31. Richard M. Nixon, "Statement on Space," Key Biscayne, FL, March 7, 1970, 2–4, folder 4702, NHRC/HQ.

32. Nixon, "Statement on Space," 2–4.

33. Space Task Group, *Post-Apollo Space Program*, 5.

34. Nixon, "Statement on Space," 3.

35. Thomas O. Paine, interview by E. M. Emme, July 9, 1970, transcript, NASA Oral History Interview, Washington, DC, folder 4186, NHRC/HQ.

36. McQuaid, "Selling the Space Age," 138.

37. See, for example, Walter R. Dornberger, "The Recoverable, Reusable Space Shuttle," *Astronautics and Aeronautics*, November 1965, 86; and Roger D. Launius, "The Strange Career of the American Spaceplane: The Long History of Wings and Wheels in Human Space Operations," *Centaurus: An International Journal of the History of Science and Its Cultural Aspects* 55, no. 4 (2013): 412–32.

38. George Mueller, "Address before the British Interplanetary Society," University College London, August 10, 1968, quoted in T. A. Heppenheimer, *Space Shuttle Decision*, 94.

39. NASA, Space Shuttle Task Group Report, "'Volume II' Desired System Characteristics," revised, June 12, 1969, in *Accessing Space*, vol. 4 of Logsdon, *Exploring the Unknown*, 206–10.

40. George Low, January 28, 1971, quoted in Scott Pace, "Engineering Design and Political Choice: The Space Shuttle, 1969–1972" (master's thesis, Massachusetts Institute of Technology, 1982), 26.

41. Dale Myers, interview by Carol Butler, August 26, 1998, transcript, Johnson Space Center Oral History Project, Houston, TX, https://historycollection.jsc.nasa.gov/JSCHistoryPortal/history/oral_histories/MyersDD/DDM_8-26-98.pdf.

42. Myers, interview, August 26, 1998.

43. Joseph P. Allen, author interview, Washington, DC, April 2, 2013.

44. Jerry Grey, *Enterprise* (New York: William Morrow, 1979), 71–72.

45. Pace, "Engineering Design and Political Choice," 30–31.

46. Grey, *Enterprise*, 73.

47. 117 Cong. Rec. S7811-12 (daily ed. May 26, 1971).

48. "Mondale: Shuttle Budget Unjustified," *Spartanburg Herald*, May 27, 1971, B3; "Mondale Will Try for Shuttle Again Next Year," *Space Business Daily*, July 26, 1971, 108, folder 7931, NHRC/HQ; J. A. Van Allen to Walter F. Mondale, May 31, 1971, folder 8271, NHRC/HQ.

49. Robert C. Cowen, "US Space Shuttle: Plan in Search of Goal," *Christian Science Monitor*, July 2, 1971, 1; Englebert Kirchner, "Sorry Virginia, There Is No Space Program," *Innovation*, April 1971, 2, 4–5, 8–9.

50. Fred L. Whipple to Clinton P. Anderson, June 15, 1971; Whipple to Anderson, June 15, 1971; Kinsey A. Anderson to Alan Cranston, June 25, 1971; all in folder 8271, NHRC/HQ.

51. Carl T. Curtis, "Space Shuttle: The Key to Our Future in Space," attachment to Alfred P. Alibrando, NASA News, July 19, 1971, folder 7931, NHRC/HQ; Howard W. Cannon, "Liftoff to Economy: The Space Shuttle," *Aerospace*, July 1971, 7.

52. "Karth Says Public Would Not Support Shuttle," *Space Business Daily*, May 19, 1971, 98, folder 1159, NHRC/HQ; "Karth Says Space Leaders Lack 'Public Interest,'" *Space Business Daily*, April 1, 1971, 161, folder 1159, NHRC/HQ; "Mondale Will Try for Shuttle Again Next Year," 108.

53. Caspar W. Weinberger, via George Shultz, to the president, memorandum, "Future of NASA," August 12, 1971, in Logsdon, *Organizing for Exploration*, 547.

54. Klaus P. Heiss and Oskar Morgenstern to James C. Fletcher, memorandum, "Factors for a Decision on a New Reusable Space Transportation System," October 28, 1971, in Logsdon, *Organizing for Exploration*, 553–54. Emphasis in original.

55. James C. Fletcher, "The Space Shuttle," November 22, 1971, in Logsdon, *Organizing for Exploration*, 558.

56. George M. Low, memorandum for the record, "Meeting with the President on January 5, 1972," January 12, 1972, in Logsdon, *Organizing for Exploration*, 558–59. All quotations are Low's account of Nixon's remarks during the conversation.

57. Richard Nixon, "Statement Announcing Decision to Proceed with Development of the Space Shuttle," January 5, 1972, *The American Presidency Project*, ed. Gerhard Peters and John T. Woolley, https://www.presidency.ucsb.edu/node/254934.

58. Nixon, "Statement Announcing Decision to Proceed."

59. Even though the moniker "space shuttle" prevailed, NASA and government officials nonetheless often would refer to it more formally as the (National) Space Transportation System.

60. Nixon, "Statement Announcing Decision to Proceed." Emphasis added.

61. Brian O'Leary, "Do We Really Want a Space Shuttle?" *New York Times*, February 16, 1972, 39; Daniel S. Greenberg, *Science and Government Report* (biweekly newsletter), June 1, 1972, 1, quoted in Roger A. Pielke Jr., "A Reappraisal of the Space Shuttle Programme," *Space Policy* 9, no. 2 (1993): 147; Bob Cromie, "Just What We Need: Another Spaceship," *Chicago Tribune*, January 8, 1972.

62. Henry M. Jackson, "Why the Space Shuttle Is Important to You," *Machinist*, February 17, 1972, quoted in 118 Cong. Rec. S3872 (daily ed. March 13, 1972).

63. "A Shuttle Boost for Florida," *Miami Herald*, January 7, 1972; "Space Shuttle 'Go' Right Decision," *Times-Picayune*, January 7, 1972.

64. See, for example, "Space Shuttle Can Be a Scientific Bridge to the Future," *Kansas City Star*, January 7, 1972; "US Stays in the Space Race," *Detroit News*, January 8, 1972.

65. Klaus P. Heiss and Oskar Morgenstern, *Mathematica Economic Analysis of the Space Shuttle System* (Princeton, NJ: Mathematica, 1972).

66. Robert F. Thompson, interview by the author, Houston, TX, June 13, 2013.

Chapter 3: Sharing the Shuttle through Discourse

1. Charles M. Lamb, "NASA, National Priorities, and the Integration of Research (draft)," September 28, 1971, ii, folder 6717, NASA Historical Reference Collection, NASA Headquarters, Washington, DC (hereafter cited as NHRC/HQ).

2. "Space Shuttle Press Briefing," Houston, TX, April 4, 1977, PC4B/1, box 4, STS General Information Subseries, Shuttle Series, Johnson Space Center History Collection, University of Houston–Clear Lake, Houston, TX (hereafter cited as JSCHC/UHCL); Roger A. Pielke Jr., "A Reappraisal of the Space Shuttle Programme," *Space Policy* 9, no. 2 (1993): 151.

3. 117 Cong. Rec. 8,151 (1971).

4. Grumman Aerospace Corporation, *Space Shuttle: The Next Logical Step*, March 1971, 19, box 4, STS General Information Subseries, Shuttle Series, JSCHC/UHCL; LF-6/ Director of Public Affairs (Brian Duff) to AD/Deputy Administrator, "Public Relations Policies," January 9, 1982, 3, folder 18175, NHRC/HQ; Jim Fraze, "Enterprise Wows Capital Citizenry," *Washington Times*, June 13, 1983; John F. Murphy to J. Scott Brownell, January 31, 1983, folder 801, NHRC/HQ.

5. Roger D. Launius, "Public Opinion Polls and Perceptions of US Human Spaceflight," *Space Policy* 19, no. 3 (2003): 167.

6. Thomas O. Paine to Clinton P. Anderson, April 3, 1970, folder 14098, NHRC/HQ; Subcommittee on NASA Oversight, "NASA Visitor Information Facilities," report submitted to US House of Representatives, Committee on Science and Astronautics, 92nd Cong., 1st sess., October 1971, committee print, 1.

7. Jon D. Miller, "Is There Public Support for Space Exploration?" *Environment*, June 1984, 30; James M. Beggs, interview by the author, Chevy Chase, MD, April 8, 2013.

8. Francis T. Hoban, *Where Do You Go after You've Been to the Moon? A Case Study of NASA's Pioneer Effort at Change* (Malabar, FL: Krieger, 1997), 5–6; "Apollo Soyuz," folder 7462, NHRC/HQ.

9. AD/Deputy Administrator (George M. Low) to RPP/Manager, Solar and Chemical Power, "NASA Publicity," June 12, 1974, folder 18175, NHRC/HQ; David Williamson Jr., "Some Thoughts on the Accurate Projection of the NASA Image," November 1, 1974, folder 18175, NHRC/HQ; NS/Director, Information Systems Division (Louis N. Lushina), to LF/Director, Public Affairs Division, "Possibility of Showing NASA Movies on Airlines," March 28, 1979, folder 18175, NHRC/HQ.

10. House Comm. on Science and Astronautics, Authorizing Appropriations to NASA, H.R. Rep. 92-143 (1971), http://www.hq.nasa.gov/office/hqlibrary/documents/045128943_1972.pdf.

11. ADA-1/Office of Policy Analysis (William B. Gevarter) to ADA/Associate Deputy Administrator, "Perception of the Space Program by the Public—Its Implications for NASA, and Some Steps It Can Take," March 27, 1975, folder 6714, NHRC/HQ.

12. Anthony J. Wiener and B. Bruce-Briggs, *Contextual Planning for NASA: A Second Workbook of Alternative Future Environments for Mission Analysis*, Interim Report II, HI-1272/3-RR (Croton-on-Hudson, NY: Hudson Institute, 1971), folder 16823, NHRC/HQ; William Overholt, Anthony J. Wiener, and Doris Yokelson, *Implications of Public Opinion for Space Program Planning, 1980–2000*, draft final report, HI-2219/2-RR (Croton-on-Hudson, NY: Hudson Institute, 1975), box 3, Space Flight Opinions and Justifications Subseries, General Reference Series, JSCHC/UHCL.

13. "Benefits of the Nation's Space Program: Summary of a Statement by Dr. Thomas O. Paine, Administrator, NASA, April 6, 1970," NASA Facts, NASA Manned Spacecraft Center, Public Affairs Office, Houston, TX, folder 4702, NHRC/HQ.

14. John Noble Wilford, "NASA, on 15th Birthday Today, Finds Itself in an Identity Crisis," *New York Times*, October 1, 1973, 70.

15. J. Kelly Beatty, "The NASA Dilemma: What Price Space?" *Sky and Telescope*, January 1978, 27, folder 6713, NHRC/HQ.

16. George M. Low to David D. Ogilvie, July 22, 1975, folder 6714, NHRC/HQ.

17. Matthew H. Hersch, *Inventing the American Astronaut* (New York: Palgrave Macmillan, 2012), 138.

18. F/Assistant Administrator for Public Affairs (John P. Donnelly) to L/Associate Administrator for External Affairs, "Space Shuttle," November 13, 1975, folder 8227, NHRC/HQ.

19. House Comm. on Science and Astronautics, Authorizing Appropriations to NASA.

20. Quoted in Gordon L. Harris, *Selling Uncle Sam* (Hicksville, NY: Exposition, 1976), 32.

21. See, for example, House Comm. on Science and Astronautics, For the Benefit of All Humankind: A Survey of the Practical Returns from Space Investment, H.R. Rep. 91-1446

(1970). The House released reports by the same name in December 1970, December 1971, October 1972, and April 1974.

22. L. B. Taylor Jr., *For All Mankind: America's Space Programs of the 1970s and Beyond* (New York: E. P. Dutton, 1974), 8; "'Now' Dictates Space Efforts, Dr. Fletcher Tells Session," *Salt Lake Tribune*, June 13, 1976, 125; Jesco von Puttkamer, "The Last Day of the Old World," *Spaceflight*, November 1979, 445.

23. NASA, "Benefits from NASA-Developed Technology," news release 72-132, July 4, 1972, folder 6990, NHRC/HQ; NASA, "Down to Earth Space Benefits Result from Moon Missions," news release 74-161, June 18, 1974, folder 6906, NHRC/HQ; NASA, "Houston Firemen Get NASA-Developed Lightweight Breathing Gear," news release 74-320, December 9, 1974, folder 6906, NHRC/HQ; NASA, "Astronaut Food Technology Applied to Meals for Elderly," news release 76-02, January 18, 1976, folder 6906, NHRC/HQ.

24. *NASA Activities*, November 1976, 28, folder 6903, NHRC/HQ; Johnson Space Center, "How to Answer Those Questions about Benefits of Space Research," *Roundup*, July 30, 1971, 3, box 2, Space Flight Opinions and Justifications Subseries, General Reference Series, JSCHC/UHCL.

25. NASA, *Space Shuttle: Emphasis for the 1970's [sic]*, NASA EP-96 (Washington, DC: US Government Printing Office, 1972), 1, box 4, STS General Information Subseries, Shuttle Series, JSCHC/UHCL; NASA, *The Space Shuttle at Work*, by Howard Allaway, NASA EP-156 (Washington, DC: US Government Printing Office, 1979), 23, quoted in Mark E. Byrnes, *Politics and Space: Image Making by NASA* (Westport, CT: Praeger, 1994), 120; Rockwell International, *Space Shuttle Transportation System* (July 1976), 3, box 4, STS General Information Subseries, Shuttle Series, JSCHC/UHCL; Grumman Aerospace Corporation, *Space Shuttle*, 12–13.

26. NASA, *Space Shuttle* (Houston: NASA Lyndon B. Johnson Space Center, February 1975), 29, folder 7854, NHRC/HQ; NASA/Rockwell, *Space Shuttle: For Down to Earth Benefits* (July 1974), folder 6906, NHRC/HQ; Rockwell, *Space Shuttle Transportation System: A Promising New Era for Earth* (September 1976), 2, box 4, STS General Information Subseries, Shuttle Series, JSCHC/UHCL.

27. NASA, *Space Shuttle* (February 1972), 13, folder 7905, NHRC/HQ; NASA, "The Space Shuttle," fact sheet, revised March 15, 1972, 3, folder 7924, NHRC/HQ; Grumman Aerospace Corporation, *Space Shuttle*, 14–15.

28. 122 Cong. Rec. S16327 (daily ed. September 21, 1976).

29. NASA Authorization for Fiscal Year 1974, *Hearing Before the Senate Comm. on Aeronautical and Space Sciences*, 93rd Cong. 218 (1973).

30. McDonnell Douglas, "Breadman . . ." (advertisement), *Forbes*, July 23, 1979, folder 11903, NHRC/HQ; Brian Duff, NASA Manned Spacecraft Center form letter response for public letters, folder 4702, NHRC/HQ; Rockwell, *Space Shuttle Transportation System: A Promising New Era for Earth*, 7; Grumman Aerospace Corporation, *Space Shuttle*, 16–17.

31. NASA, *Space Shuttle* (Washington, DC: US Government Printing Office, 1972), box 4, STS General Information Subseries, Shuttle Series, JSCHC/UHCL; NASA Authorization for Fiscal Year 1974, *Hearing Before the Senate Comm. on Aeronautical and Space Sciences*, 93rd Cong. 216 (1973); NASA, *Space Shuttle: Emphasis for the 1970's*, 8; Rockwell International, *Space Shuttle Transportation System* (July 1976), 12.

32. AD/Deputy Administrator (George M. Low) to distribution, "Meetings with Non-aerospace Business and Community Leaders," April 15, 1975, folder 4159, NHRC/HQ; George M. Low to Daniel J. Fink, December 31, 1974, folder 6714, NHRC/HQ; George M. Low to Glenn S. Dumke, June 9, 1975, folder 6714, NHRC/HQ; "Naugle Speaks on Space Program at FASST/White House Youth Conference," *NASA Activities*, April 1975, 18, "FASST—publicity" folder, box 7, Alan Ladwig unprocessed papers, NHRC/HQ.

33. K/Assistant Administrator for Industry Affairs and Technology Utilization (Edward Z. Gray) to A/Administrator, "Phone Call to Mr. Robert Anderson, Rockwell," December 30, 1974, folder 5005, NHRC/HQ; K/Assistant Administrator for Industry Affairs and Technology Utilization (Edward Z. Gray) to A/Administrator and AD/Deputy Administrator, "Organization of a National Space Organization," October 30, 1973, folder 5005, NHRC/HQ; K/Assistant Administrator for Industry Affairs and Technology Utilization (Edward Z. Gray) to A/Administrator, "Meeting with National Space Club Representatives—10:00 A.M., Friday, December 7, 1973," December 5, 1973, folder 5005, NHRC/HQ.

34. James C. Fletcher to Wernher von Braun, April 18, 1975, folder 5005, NHRC/HQ.

35. Hugh Downs, "Testimony of Mr. Hugh Downs, Vice President, National Space Institute, before the Subcommittee on Aerospace Technology and National Needs," folder 498, NHRC/HQ; National Space Institute, *The National Space Institute* (Arlington, VA: National Space Institute, n.d.), folder 5005, NHRC/HQ; Frederick I. Ordway III, "Wernher von Braun and the National Space Institute," *Ad Astra*, November/December 1994, 31; Wernher von Braun, "For Space Buffs—National Space Institute: You Can Join," *Popular Science*, May 1976, 72; National Space Institute, "NSI's Executive Director Testifies before House Subcommittee on NASA Budget and the US Space Program," news release, "Miscellaneous Space Surveys" folder, box 16, Alan Ladwig unprocessed papers, NHRC/HQ.

36. Krafft A. Ehricke, "Extraterrestrial Imperative," *Bulletin of the Atomic Scientists*, November 1971, 18–26; Gerald K. O'Neill, "The Colonization of Space," *Physics Today*, September 1974, 32–40.

37. Daniel A. Petitpas, "Nation Needs Another Great Goal," *Boston Globe*, July 31, 1977.

38. Hans Mark to Jim Fletcher, May 9, 1976, folder 4215, NHRC/HQ; F/Acting Assistant Administrator for Public Affairs (Robert J. Shafer) to Ames Research Center/Hans Mark, "Science Fiction Audiences," May 17, 1976, folder 4215, NHRC/HQ. Fletcher's sentiment appears as a handwritten note on the memorandum.

39. NASA, *Space Resources and Space Settlement*, NASA SP-428 (Washington, DC: US

Government Printing Office, 1979); NASA, *The Shuttle Era* (Houston: Lyndon B. Johnson Space Center, December 1977), 3, box 4, STS General Information Subseries, Shuttle Series, JSCHC/UHCL.

40. Rockwell, *Space Shuttle Transportation System: A Promising New Era for Earth*, 24.

41. Peter Cobun, "NASA 'Old Guard' Bristling," *Birmingham News*, October 5, 1976. "Star 'Trekkies' are Briefed on NASA Goals," *NASA Activities*, March 1976, 27; folder 8935, NHRC/HQ.

42. David Brandt-Erichsen, "The L-5 Society," *Ad Astra*, November/December 1994, 35.

43. Henry W. Pierce, "NASA Chief Predicts People Space Colonies," *Pittsburgh Post-Gazette*, September 28, 1974; Homer E. Newell, *Beyond the Atmosphere: Early Years of Space Science* (Washington, DC: US Government Printing Office, 1980), 397; "'Now' Dictates Space Efforts, Dr. Fletcher Tells Session," *Salt Lake Tribune*, June 13, 1976, 125.

44. Barbara Hubbard to Thomas Paine, June 4, 1968; Homer E. Newell to Waldo E. Smith, June 10, 1968; Hubbard to George Lowe [*sic*], March 25, 1971. All documents located in folder 15114, NHRC/HQ.

45. Barbara Marx Hubbard, *The Hunger of Eve* (Harrisburg, PA: Stackpole, 1976), 143–49; George M. Low to Hubbard, April 5, 1971; Associate Administrator for Manned Space Flight (Dale D. Myers) to George M. Low, "Barbara Hubbard's Committee for the Future," July 20, 1972; John E. Naugle, "Meeting with Barbara Marx Hubbard (Chairman, The Committee for the Future, Inc.)," memorandum for file, January 24, 1975. Fletcher ultimately denied the request because of security and logistical issues around the shuttle's use of the facility. See Fletcher to Hubbard, May 29, 1975. All documents located in folder 15114, NHRC/HQ.

46. Associated Press, "And Now, a Real Space Shuttle Named 'Enterprise,'" *Washington Post*, September 9, 1976, A2. Brian Woods posits that Ford sided with the Trekkies and endorsed the name *Enterprise* because he believed it would help him in the upcoming presidential election. See Brian Woods, "Artifacts, Revolutionaries, and Bureaucrats: The Sociotechnical Shaping of NASA's Space Shuttle" (PhD diss., University of Edinburgh, 1998), 223.

47. Jonathan Eberhart, "Space Shuttle: An Enterprising Debut," *Science News*, September 25, 1976, 199; Associated Press, "And Now, a Real Space Shuttle Named 'Enterprise,'" A2.

48. Mary Manning, letter to the editor, *Los Angeles Times*, January 20, 1973.

49. NASA, *Man in Space: Space in the Seventies*, by Walter Froehlich, NASA EP-81 (Washington, DC: US Government Printing Office, 1971), 11, quoted in Byrnes, *Politics and Space*, 117.

50. NASA, *Shuttle Era*, 6; Dick Baumbach, "Get Away! Vacation in Space No Flight of Fancy," *Today*, February 17, 1980, 12D, folder 8164, NHRC/HQ; William K. Stevens, "New Generation of Astronauts Poised for Shuttle Era," *New York Times*, April 6, 1981, A1, B10; NASA, "Space Shuttle: NASA's Answer to Operations in Near-Earth Orbit," NASA Facts,

NF-144, October 1985, 5, folder 7905, NHRC/HQ; NASA, *Space Shuttle: Emphasis for the 1970's*, 6; "Space Shuttle Good Business: NASA Chief," *Chicago Tribune*, March 28, 1982.

51. "What? Space Shuttles!" attachment to O. B. Lloyd Jr., letter to the editor, March 30, 1970, folder 18179, NHRC/HQ; NASA, *Space Shuttle: Emphasis for the 1970's*, 6.

52. Jon Schneeberger, Ken Dallison, and Rick Gore, "When the Space Shuttle Finally Flies," *National Geographic Magazine*, March 1981, 317; Pat Legan to Information Center, NASA, Johnson Space Center, January 4, 1979, box 4, Public Affairs Subseries, Center Series, JSCHC/UHCL.

53. NASA, *Space Shuttle* (February 1975), 1; NASA, *Shuttle Era*, 3.

54. Grumman Aerospace Corporation, *Space Shuttle*. The term "space truck" would be used by various entities within and outside of NASA through the first several years of the shuttle's operations; NASA/TWA, "Don't Miss the Excitement at Kennedy Space Center" (leaflet), [1975?], folder 14098, NHRC/HQ.

55. NASA, *Space Shuttle: Emphasis for the 1970's*.

56. Rockwell, *Space Shuttle: A Promising New Era for Earth* (February 1977), box 4, STS General Information Subseries, Shuttle Series, JSCHC/UHCL; "Fletcher Says Shuttle Marks New Era," *Defense/Space Business Daily*, September 21, 1976, 110, folder 4215, NHRC/HQ.

57. *America's Spaceport*, n.d., 22, folder 4636, NHRC/HQ.

58. Joseph B. Treaster, "Nation Reacts with Cheers and 'We're No. 1 Again,'" *New York Times*, April 15, 1981, A24.

59. Henry S. F. Cooper, "Shuttle-I," *New Yorker*, February 9, 1981, 76; Alan Ladwig, interview by the author, Washington, DC, January 31, 2013.

60. Ronald Kotulak and Jon Van, "Our Future Cities May Be 'Out of This World,'" *Chicago Tribune*, April 9, 1981.

61. James C. Fletcher to J. Carter Brown, October 3, 1975, folder 4159, NHRC/HQ.

62. James M. Beggs, "Suggested Remarks: SCA/Enterprise Ceremony," Dulles International Airport, June 12, 1983, NHRC online, https://historydms.hq.nasa.gov/sites/default/files/DMS/e000042016.pdf; Ronald Reagan, "Statement on the Landing of the Space Shuttle *Columbia* Following Its Inaugural Flight, April 14, 1981," *Public Papers of the Presidents: Ronald Reagan, 1981* (Washington, DC: US Government Printing Office, 1982), 353; Jacquie Harbour, "Abrahamson Addresses Texas Newsmen Meeting," *Citizen*, March 28, 1982, folder 11, NHRC/HQ; "The Space Shuttle's Abrahamson: It's [*sic*] Potential Is Immense," *Government Executive*, July/August 1983, 11.

63. James A. Abrahamson, "Can the Space Shuttle Compete?," remarks to the Aviation/Space Writers Association, April 12, 1983, 13–14, NHRC online, https://historydms.hq.nasa.gov/sites/default/files/DMS/e000043021.pdf; Ronald Reagan, "Address before a Joint Session of the Congress on the Program for Economic Recovery, April 28, 1981," *Public Papers of the Presidents: Ronald Reagan, 1981* (Washington, DC: US Government Printing Office, 1982), 394.

64. NASA John F. Kennedy Space Center, *Welcome to Your Spaceport* [1977?], folder 4636, NHRC/HQ; Michael Mecham, "Thousands Will Make Space Flight by 2000," *USA Today*, February 3, 1984, 10A; Thomas O'Toole, "Shuttle's 'Cleanest Mission' Ends with Night Landing," *Washington Post*, September 6, 1983, A3.

65. James B. Odom, interview by Rebecca Wright, July 20, 2010, Huntsville, AL, transcript, NASA STS Recordation Oral History Project, Houston, TX, accessed December 17, 2014, https://web.archive.org/web/20130317233624/http://www.jsc.nasa.gov/history/oral_histories/STS-R/OdomJB/OdomJB_7-20-10.pdf.

66. See, for example, Peter J. Sacchetti to Sirs, February 23, 1980; Sam Huston to Christopher C. Kraft Jr., March 12, 1981; and Daniel B. Littmann to Kraft, July 19, 1981; all in box 4, Public Affairs Subseries, Center Series, JSCHC/UHCL.

67. Lee Hickling, "GAO Criticizes Shuttle—From Tires to Cost," *Today*, June 4, 1977, 8A, folder 7958, NHRC/HQ.

68. "The Plane That Won't Fly," *Newsweek*, May 5, 1980, 94–96; Arlen J. Large, "America's Space Shuttle Lemon," *Newsweek*, January 31, 1980, 14; Joel S. Hirschhorn, "Aluminum Dumbo," *Washington Post*, March 28, 1980, A19; and Gregg Easterbrook, "The Spruce Goose of Outer Space," *Washington Monthly*, April 1980; Robert F. Thompson, interview by the author, Houston, TX, June 13, 2013.

69. "US Picks a Name for Shuttle, Can't Predict Launch," *Wall Street Journal*, January 26, 1979, 1. Skylab fell to Earth on July 11, 1979, landing in Western Australia.

70. Brian Duff, interview by John Mauer, May 24, 1989, transcript, box 12, folder 16, Glennan-Webb-Seamans Project, National Air and Space Museum Archives, Smithsonian Institution, Washington, DC; Duff, interview by Mauer, May 1, 1989, transcript, box 12, folder 15, Glennan-Webb-Seamans Project, National Air and Space Museum Archives.

71. LF-6/Director of Public Affairs to A/Administrator and AD/Deputy Administrator, "Sustained Information/Education Program on the Shuttle Transportation System," June 23, 1980, folder 18175, NHRC/HQ.

72. Duff, interview by Mauer, May 1, 1989.

73. Duff, interview by Mauer, May 1, 1989; Malcolm W. Browne, "News Media Swarm to Shuttle Site, Giving Launching a Carnival Air," *New York Times*, April 9, 1981, B17.

74. Lisa Malone, interview by the author, Arlington, VA, April 17, 2013; LF-6/Director of Public Affairs (Brian Duff) to AD/Deputy Administrator, "Public Relations Policies," January 9, 1982, 2, folder 18175, NHRC/HQ.

75. See, for example, Lee Dembart, "Pulling Success Out of the Hat, NASA Remains a 'Can-Do' Government Agency," *Los Angeles Times*, April 22, 1984, and "Columbia Aloft," *Washington Star*, April 14, 1981.

76. LF-6/Director of Public Affairs (Brian Duff) to AD/Deputy Administrator, "Public Relations Policies," January 9, 1982, 2, folder 18175, NHRC/HQ.

77. NASA, *How to Cover the First Space Shuttle Mission*, Space Transportation System

Briefing Series, no. 7 (January 12, 1981), folder 18179, NHRC/HQ; NASA, *Space Shuttle News Media Handbook* (February 1982), box 4, STS General Information Subseries, Shuttle Series, JSCHC/UHCL.

78. NASA, "Worldwide Telephone Service Set Up for STS-6 Mission," news release 83-41, March 28, 1983, folder 5005, NHRC/HQ; John F. Murphy to George Carvalho, July 20, 1983, folder 801, NHRC/HQ; Henry S. F. Cooper, *Before Lift-Off: The Making of a Space Shuttle Crew* (Baltimore, MD: Johns Hopkins University Press, 1987).

79. Rockwell, *Shuttle* (July 1982), box 4, STS General Information Subseries, Shuttle Series, JSCHC/UHCL, 2.

80. NASA, Lyndon B. Johnson Space Center, *The Shuttle Era*, JSC-12599 (Houston, TX: NASA, 1983), 2.

81. Thomas O'Toole, "Challenger Touches Down; Flight Is Called Best So Far," *Washington Post*, April 10, 1983, A7.

82. Malone, interview by the author, April 17, 2013.

83. See, for example, NASA, Division of Public Affairs, *Aboard the Space Shuttle*, by Florence S. Steinberg, EP-169 (Washington, DC: NASA, 1980), folder 8163, NHRC/HQ, and NASA, John F. Kennedy Space Center, *Spaceage Foods for Space Travelers*, KSC 85-80 (Kennedy Space Center, FL: NASA, July 1981), "Space Food" folder, box 14, Alan Ladwig unprocessed papers, NHRC/HQ.

84. Department of Housing and Urban Development, and Certain Independent Agencies Appropriations for 1986, *Hearing Before the House Committee on Appropriations, Subcommittee on HUD-Independent Agencies*, 98th Cong. 39, A8 (1983).

Chapter 4: Inviting Virtual Participation

1. James W. McCulla, "Guest Activities in the Shuttle Era," March 31, 1982, attached to LFF 3/ Chief, Public Services Branch to distribution, "Revising NASA Guest Activities in the Shuttle Era," April 1, 1982, NASA Historical Reference Collection, NASA Headquarters, Washington, DC (hereafter cited as NHRC/HQ).

2. David E. Nye, *American Technological Sublime* (Cambridge: MIT Press, 1994); James M. Beggs, interview by the author, April 8, 2013; Lisa Malone, interview by the author, April 17, 2013.

3. James Hartsfield, interview by the author, Houston, TX, June 12, 2013.

4. Thomas O'Toole, "Space Shuttle Orbiter Shown for First Time in California," *Washington Post*, September 18, 1976, A2.

5. NASA, "Public Affairs Plan: Approach and Landing Test, January 1977 through March 1978," January 26, 1977, box 2, ALT Documents Subseries, Shuttle Series, Johnson Space Center History Collection, University of Houston–Clear Lake, Houston, TX (hereafter cited as JSCHC/UHCL); NASA, *Astronautics and Aeronautics, 1978: A Chronology*, NASA SP-4023 (Washington, DC: US Government Printing Office, 1986), 59–60.

6. MH-7/Director, Space Shuttle Program (Myron S. Malkin) to M/Associate Administrator for Space Transportation Systems, "Feasibility of Transporting OV-101 to the Paris Air Show," November 24, 1978, folder 8009, NHRC/HQ.

7. Brian Duff, interview by John Mauer, May 24, 1989, transcript, box 12, folder 16, Glennan-Webb-Seamans Project, National Air and Space Museum Archives, Smithsonian Institution, Washington, DC. See also James M. Beggs to Charles McC. Mathias Jr., February 18, 1983, folder 8009, NHRC/HQ.

8. Paul Lewis, "Paris Bars a Flight by Shuttle over Seine," *New York Times*, May 22, 1983; James Abrahamson, interview by the author, Washington, DC, June 19, 2013; Louis A. Parker, interview by Rebecca Wright, Houston, TX, December 6, 2001, transcript, NASA Johnson Space Center Oral History Project, Houston TX, April 30, 2023, https://historycollection.jsc.nasa.gov/JSCHistoryPortal/history/oral_histories/ParkerLA/ParkerLA_12-6-11.pdf.

9. Parker, interview by Wright, December 6, 2001; James M. Beggs, "Suggested Remarks: SCA/Enterprise Ceremony," Dulles International Airport, June 12, 1983, NHRC online, https://historydms.hq.nasa.gov/sites/default/files/DMS/e000042016.pdf; Jim Fraze, "Enterprise Wows Capital Citizenry," *Washington Times*, June 13, 1983; John Burgess, "Earthlings Applaud as Enterprise Makes Orbit above Beltway," *Washington Post*, June 13, 1983, B1, B6–B7.

10. NASA, "Shuttle Orbiter Enterprise to be Shown at New Orleans Exposition," news release 83-188, December 1, 1983, folder 8011, NHRC/HQ.

11. Plans to send *Enterprise* to the 1982 world's fair in Knoxville, Tennessee, were scuttled in part due to the inability of a barge carrying the orbiter to pass under several of the bridges spanning the route the orbiter would have to travel. Peter Adams, "Enterprise Won't Make World's Fair Date," *Today*, April 29, 1982, folder 8009, NHRC/HQ.

12. Parker, interview by Wright, December 6, 2001; "Enterprise Popular Attraction at World's Fair," *NASA Activities*, February 1985, 13, folder 8009, NHRC/HQ.

13. George M. Low (for James C. Fletcher) to Spiro T. Agnew, March 30, 1973, folder 14098, NHRC/HQ; James C. Fletcher to Olin E. Teague, September 25, 1974, folder 14098, NHRC/HQ; Gina Seay, "Tours Share Space with the World," *Houston Post*, May 30, 1986, 1E, 14E.

14. National Space Institute, advertisement, *Space World*, February 1985, quoted in Nye, *American Technological Sublime*, 248.

15. Malcolm W. Browne, "News Media Swarm to Shuttle Site, Giving Launching a Carnival Air," *New York Times*, April 9, 1981, B17; McCulla, "Guest Activities in the Shuttle Era."

16. "'Yeeeow!' and 'Doggone!' Are Shouted on Beaches as Crowds Watch Liftoff," *New York Times*, April 13, 1981, A17; "Pool Status #2 (final)," April 12, 1981, folder 8314, NHRC/HQ.

17. Julian Scheer to distribution, memorandum, "Invitations to View Launch of Apollo 5," December 18, 1967; Scheer to distribution, memorandum, "Protocol Plans for Apollo Manned Missions," June 17, 1968, both memoranda in folder 1904, NHRC/HQ.

18. NASA, "Public Affairs Plan, Apollo 9," February 1969, folder 18179, NHRC/HQ; McCulla, "Guest Activities in the Shuttle Era"; "The Space Shuttle's Abrahamson: It's [sic] Potential Is Immense," *Government Executive*, July/August 1983, 11; Alan Ladwig, interview by the author, January 31, 2013.

19. See, for example, William Overholt, Anthony J. Wiener, and Doris Yokelson, *Implications of Public Opinion for Space Program Planning, 1980–2000*, draft final report, HI-2219/2-RR (Croton-on-Hudson, NY: Hudson Institute, 1975); and Jon D. Miller, "The Information Needs of the Public Concerning Space Exploration: A Special Report for the National Aeronautics and Space Administration," July 20, 1982, folder 6720, NHRC/HQ.

20. Duff, interview by Mauer, May 24, 1989; Josie A. Soper, interview by Rebecca Wright, April 19, 2006, transcript, NASA Headquarters Administrator Project, NASA Headquarters, Washington, DC, https://historycollection.jsc.nasa.gov/JSCHistoryPortal/history/oral_histories/NASA_HQ/Administrators/SoperJA/SoperJA_4-19-06.pdf.

21. Thomas O'Toole, "Shuttle's 'Cleanest Mission' Ends with Night Landing," *Washington Post*. September 6, 1983, A3; Duff, interview by Mauer, May 24, 1989; Beth Beck, interview by the author, June 18, 2013.

22. NASA, "Space Transportation System Public Affairs Plan: First Space Shuttle Mission, STS-1," September 1980, 5-1, Internet Archive, https://archive.org/details/nasa_techdoc_19900067371.

23. Carla Hall, *Washington Post*, December 1, 1982, final edition, C7, folder 5308, NHRC/HQ.

24. Smithsonian Institution, "Space Shuttle Art Exhibition to Go on View at the National Air and Space Museum," news release 398-82, November 12, 1982; NASA, *Artists and the Space Shuttle* (N.p.: NASA, n.d.); Owen McNally, "Depicting US Push into Space," *Hartford (CT) Courant*, April 18, 1984; NASA, "Composer/Musician to Cover Shuttle Mission under NASA Art Program," news release 88-128, September 19, 1988; all in folder 5308, NHRC/HQ.

25. NASA, *A Meeting with the Universe*, ed. Bevan M. French and Stephen P. Moran, EP-177 (Washington, DC: NASA, 1981), iv.

26. Brian Duff, interview by John Mauer, April 24, 1989, transcript, Glennan-Webb-Seamans Project, box 12, folder 13, National Air and Space Museum Archives.

27. Robert N. Jacobs, email message to the author, June 6, 2014; NASA, "Miniature TV Camera to Be Used on Fifth Shuttle Mission," news release 82-135, September 15, 1982, folder 7038, NHRC/HQ; FD-1/Deputy Assistant Administrator for Television (Robert J. Shafer) to L/Associate Administrator for External Affairs, "Space Shuttle Television," December 15, 1975, folder 8227, NHRC/HQ.

28. William Readdy, interview by the author, Arlington, VA, June 27, 2014; Robert N. Jacobs, interview by the author, January 24, 2013.

29. Brian Duff, "*The Dream Is Alive*—The Story behind a Historic Film," *NASM Research Report*, 1985, 195–201, folder 15429, NHRC/HQ; Joseph P. Allen, interview by Jennifer Ross-Nazzal, Washington, DC, March 16, 2004, transcript, Johnson Space Center Oral History Project, Houston, TX, https://historycollection.jsc.nasa.gov/JSCHistoryPortal/history/oral_histories/AllenJP/AllenJP_3-16-04.htm.

30. Duff, "*Dream Is Alive*," 198.

31. John F. Murphy to Thad Cochran, April 14, 1983, folder 6414, NHRC/HQ.

32. AT&T, "During the Next 6 Days, You Can Make an Extraordinary Long Distance Call" (advertisement), publication not indicated, 1984, folder 5005, NHRC/HQ; Mark R. Chartrand, "National Space Institute: The Middle Years, 1980–1984," *Ad Astra*, November/December 1994, 32–33, folder 5006, NHRC/HQ; National Space Society/Rockwell, "Dial-A-Shuttle: The Ultimate Long Distance Call" (advertisement), *Ad Astra*, May/June 1995, back cover, folder 5006, NHRC/HQ.

33. George W. S. Abbey, interview by the author, Houston, TX, June 12, 2013; Beggs, interview by the author, April 8, 2013; LFF/Chief of Special Events Branch (Gene Marianetti) to L/Associate Administrator for External Relations, "Year End Report of Astronaut Appearances," January 25, 1984, folder 8929, NHRC/HQ.

34. M/Associate Administrator for Space Flight (Jesse W. Moore) to L/Associate Administrator for External Relations, "Sesame Street Television Production," November 22, 1985, folder 19758, NHRC/HQ.

35. NASA, "Ham Radio to Fly on STS-9," news release 83-90, June 3, 1983, folder 16907, NHRC/HQ.

36. LF/Director, Public Affairs Division (Shirley M. Green) to M/Associate Administrator for Space Flight, February 18, 1986, folder 1260, NHRC/HQ; Patricia Palazzolo, "Launching Dreams: The Long-Term Impact of SAREX and ARISS on Student Achievement," n.d., http://www.amsat.org/amsat/ariss/Presentations/AMSAT%20Symposiums/2007%20Symposium/Launching%20Dreams.pdf; NASA, *Wings in Orbit: Scientific and Engineering Legacies of the Space Shuttle*, ed. Wayne Hale, Helen Lane, Gail Chapline, and Kamlesh Lula, NASA SP-2010-3409 (Washington, DC: US Government Printing Office, 2010), 471–73.

37. Carolyn Sumners, interview by the author, April 26, 2013; "Toys in Space: Results of the Space Shuttle Mission STS-51D," n.d., box 23, General History V Subseries, General Reference Series, JSCHC/UHCL.

38. Robert Reinhold, "Reagan Warns Schools Are Failing to Meet Science and Math Needs," *New York Times*, May 13, 1982, B18; National Space Institute, "Young Astronauts Program to Be Formed," news release, June 20, 1984, folder 8926, NHRC/HQ; Ellen I. Kelly, "The Young Astronauts," *AIA Aerospace*, Summer 1985, 14–16.

39. Jack Anderson, "Dear Educator" letter, October 24, 1984, folder 8926, NHRC/HQ; Jack Anderson to James M. Beggs, November 21, 1985, "Young Astronauts" folder, box 4, Alan Ladwig unprocessed papers, NHRC/HQ; XE/Director, Educational Affairs Division (Robert W. Brown) to XE/Staff, "Update on Young Astronaut Council (YAC)," September 21, 1987, folder 8926, NHRC/HQ.

40. Lawrence F. Herbolsheimer to James. M. Beggs, "Young Astronauts Program," April 5, 1985; Beggs to Herbolsheimer, April 25, 1985; both in "Young Astronauts" folder, box 4, Alan Ladwig unprocessed papers, NHRC/HQ.

41. Susan Alsup, "Move Over Monkey Bars," *Space News Roundup*, September 21, 1990, 3.

42. NASA Lewis Research Center, *Launching a Dream: A Teachers Guide to a Simulated Space Shuttle Mission* (Washington, DC: US Government Printing Office, 1988)., folder 7906, NHRC/HQ.

Chapter 5: Using the Space Truck

1. Ronald Reagan: "Fact Sheet Outlining United States Space Policy," July 4, 1982, Ronald Reagan Presidential Library and Museum, https://www.reaganlibrary.gov/archives/speech/fact-sheet-outlining-united-states-space-policy.

2. National Aeronautics and Space Act of 1958, Pub. L. No. 85-568 72 Stat. 426-2 (1958).

3. "NASA Selling Space Medicine Idea," *Today* (Florida), May 25, 1978, 8A, folder 8189, NASA Historical Reference Collection, NASA Headquarters, Washington, DC (hereafter cited as NHRC/HQ).

4. Rockwell International, *Space Shuttle: An Opportunity for University Research* (Rockwell, n.d.), 7-8, folder 19815, NHRC/HQ; NASA, *The Role of the Payload Specialist in the Space Transportation System for NASA or NASA-Related Payloads* (NASA, n.d.), folder 8971, NHRC/HQ; Payload Specialists for NASA or NASA-Related Payloads, 43 Fed. Reg. 9790 (March 10, 1978) (amending 14 C.F.R. pt. 1214); NASA, *Space Shuttle* (Washington, DC: US Government Printing Office, 1972).

5. National Space Institute, "LDEF for New Experimenters," *Newsletter of the National Space Institute*, August 1976, 1-2, folder 1260, NHRC/HQ; Leonard W. David, "Student Experimentation via the Shuttle (An Overview)," paper presented at the 85th annual conference of the American Society for Engineering Education, University of North Dakota, June 27-30, 1977, 8-9, folder 19814, NHRC/HQ; NASA, Langley Research Center, *69 Months in Space: A History of the First LDEF (Long Duration Exposure Facility)*, NP-149 (Hampton, VA: NASA, 1992), inside cover.

6. Associate Administrator for Manned Space Flight (Dale D. Myers) to James C. Fletcher, February 4, 1972, NHRC/HQ.

7. "How NASA Went into the Freight Business," *Business Week*, August 22, 1977; Kevin Anderson, "Shuttle Shoots for Customers and Profits," *Washington Post*, November 10, 1982, 1B.

8. Jim Maloney, "'Interest Expressed' in Shuttle as Commercial Spaceliner," *Houston Post*, November 3, 1978, folder 8131, NHRC/HQ.

9. James M. Beggs, interview by the author, April 8, 2013.

10. Mark E. Byrnes, *Politics and Space: Image Making by NASA* (Westport, CT: Praeger, 1994), 125; Ronald Reagan, "Address before a Joint Session of the Congress on the State of the Union, January 25, 1984," *Public Papers of the Presidents: Ronald Reagan, 1984*, bk. 1 (Washington, DC: US Government Printing Office, 1986–1987), 90.

11. Jon M. Smith, interview by the author, Houston, TX, June 10, 2013; "How NASA Went into the Freight Business"; Erwin J. Bulban, "Non-aerospace Firms Eyed for Shuttle," *Aviation Week & Space Technology*, April 5, 1976, 52–55.

12. James Abrahamson, interview by the author, June 19, 2013; "NASA's Tough Job of Selling Its Space Shuttle," *Business Week*, December 25, 1978; Charlie Peek, "Space Agency Sheds 'Gee-Whiz' Approach," *Today*, February 11, 1979, 22C, folder 8131, NHRC/HQ.

13. NASA, *Ready to Serve You in Space*, n.d., folder 19815, NHRC/HQ; McDonnell Douglas, "Room for Rent" (advertisement), *Wall Street Journal*, May 3, 1978; "NASA's Tough Job of Selling Its Space Shuttle"; Rockwell International, "Rockwell Can Fit You In" (advertisement), 1978, folder 11903, NHRC/HQ; Rockwell International, *Integrating Payloads into Space Shuttle* (Downey, CA: Rockwell, 1976), box 14, Alan Ladwig unprocessed papers, NHRC/HQ; Rockwell, *Space Shuttle: A Promising New Era for Earth* (February 1977), box 4, STS General Information Subseries, Shuttle Series, Johnson Space Center History Collection, University of Houston–Clear Lake, Houston, TX (hereafter cited as JSCHC/UHCL).

14. See Thomas O'Toole, "'We Deliver' Shuttle Comes Home," *Washington Post*, November 17, 1982; NASA, *We Deliver* (Washington, DC: US Government Printing Office, 1983), folder 8084, NHRC/HQ.

15. See, for example, the French-language version: NASA, *We Deliver: Comptez sur Nous!* (Washington, DC: US Government Printing Office, 1983); "Space Shuttle Marketing Effort Planned at Paris Air Show," *Defense Daily*, May 17, 1983, 94; both in folder 8084, NHRC/HQ.

16. NASA, "The Space Transportation System Marketing Plan," June 1984, folder 8085, NHRC/HQ.

17. Reimbursement for Shuttle Services Provided to Non-US Government Users, 42 Fed. Reg. 3829 (January 21, 1977) (amending 14 C.F.R. pt. 1214); NASA, *Space Transportation Operations Information* (Washington, DC: NASA, [1981?]), folder 8183, NHRC/HQ; John Noble Wilford, "'Bargain' Prices Set for Space Shuttle," January 13, 1977, 21, folder 8084, NHRC/HQ; Abrahamson, interview by the author, June 19, 2013.

18. Charles D. Walker, interview by Jennifer Ross-Nazzal, Washington, DC, November 19, 2004, transcript, Johnson Space Center Oral History Project, Houston TX, https://historycollection.jsc.nasa.gov/JSCHistoryPortal/history/oral_histories/WalkerCD/WalkerCD_11-19-04.htm; NASA, "NASA-McDonnell Douglas Sign Joint Endeavor to

Stimulate Commercialization of Space," news release 80-8, January 25, 1980, folder 8189, NHRC/HQ; NASA, Office of Space Science and Applications, Microgravity Science and Applications Division, *Microgravity . . . A New Tool for Basic and Applied Research in Space*, EP-212 (Washington, DC: US Government Printing Office, n.d.), folder 7906, NHRC/HQ.

19. NASA, "Space Transportation System Marketing Plan."

20. NASA, "ASTP Experiment May Resolve the Value of Electrophoresis in Space," news release 75-178, n.d., folder 8189, NHRC/HQ; Walker, interview by Ross-Nazzal, November 19, 2004.

21. Arlen J. Large, "Space Shuttle Plans Free Rides for Firm to Do Tests in Orbit," *Wall Street Journal*, February 1, 1982, 18; "Shuttle Crystal-Growing Experiment 'Unqualified Success': 3M Co.," *Aerospace Daily*, December 5, 1984, 174; Michael Schrage, "Business Use of Shuttle Slower Than Expected," *Washington Post*, March 7, 1984, D16; Eric Gelman, "Starship 'Free Enterprise,'" *Newsweek*, September 17, 1984, 63. See also NASA, Marshall Space Flight Center, *Monodisperse Latex Reactor (MLR): A Materials Processing Space Shuttle Mid-deck Payload*, by Dale M. Kornfeld, TM-86487 (Washington, DC: US Government Printing Office, 1985).

22. James Fisher, "Space: The Final Frontier for Coke, Pepsi," *Orlando Sentinel*, July 6, 1986; NASA, "NASA to Test Dispensing Technology in Space," news release 85-97, June 25, 1985, folder 5718, NHRC/HQ.

23. AD/Deputy Administrator (George M. Low) to M/Associate Administrator for Space Flight, "Space Shuttle Operations," December 30, 1975, NHRC/HQ; Abrahamson, interview by the author, June 19, 2013.

24. NASA, "NASA Announces New Shuttle Launch Prices for 1985–88," news release 82-94, June 15, 1982, folder 8084, NHRC/HQ; Craig Covault, "NASA Curtails Shuttle Flights," *Aviation Week & Space Technology*, June 21, 1982, 16–18.

25. "DOT Wants Shuttle Prices Based on Full Cost Recovery," *Defense Daily*, March 12, 1985, 62; "NASA-DOT Conflict Emerges Over Shuttle Pricing Policy," *Aerospace Daily*, March 8, 1985; Elizabeth Tucker, "Space Shuttle Pricing Debated," *Washington Post*, June 20, 1984, F1; Mary Belcher, "NASA to Sell Shuttle Space by Auction," *USA Today*, August 2, 1985, 10C.

26. "NASA's Tough Job of Selling Its Space Shuttle"; John F. Yardley, interview by Summer Chick Bergen, Saint Louis, MO, June 30, 1998, transcript, Johnson Space Center Oral History Project, Houston, TX, https://historycollection.jsc.nasa.gov/JSCHistoryPortal/history/oral_histories/YardleyJF/JFY_6-30-98.pdf; Charles D. Walker, interview by Sandra Johnson, Springfield, VA, November 7, 2006, transcript, Johnson Space Center Oral History Project, Houston TX, https://historycollection.jsc.nasa.gov/JSCHistoryPortal/history/oral_histories/WalkerCD/WalkerCD_11-7-06.pdf.

27. Beggs, interview by the author, April 8, 2013.

28. Gil Moore, "Remarks at the Funeral of NASA's Get Away Special Program: An

Open Letter to Dr. Michael Griffin," March 20, 2005, SpaceRef, https://spaceref.com/uncategorized/remarks-at-the-funeral-of-nasas-get-away-special-program-an-open-letter-to-dr-michael-griffin/; M/Associate Administrator for Space Flight (John F. Yardley) to distribution, "Small Self-Contained Payload Concepts," with attachment, July 28, 1977, folder 8184, NHRC/HQ.

29. NASA, Office of Space Flight, Customer Services Division, *Get-Away Specials* (Washington, DC: US Government Printing Office, n.d.), folder 8183, NHRC/HQ; James A. Abrahamson, "Can the Space Shuttle Compete?," remarks to the Aviation/Space Writers Association, April 12, 1983, 13–14, NHRC online, https://historydms.hq.nasa.gov/sites/default/files/DMS/e000043021.pdf; NASA, Goddard Space Flight Center, Special Payloads Division, *Get Away Special . . . The First Ten Years* (Greenbelt, MD: NASA, 1989), 3, folder 8182, NHRC/HQ; Ike Flores, "Here's a GAS of a Cargo Flight," *Washington Star*, May 14, 1978, A-14. While NASA opened the opportunity for space experimentation to a variety of people, the $3,000–$10,000 price tag plus the additional costs of developing experiments kept the prospects of participation beyond the reach of many.

30. NASA, *Get Away Special . . . The First Ten Years*, 3; NASA, "Space Transportation System: Small Self-Contained Payload Program," September 1977, folder 8182, NHRC/HQ; Alan Ladwig, interview by the author, January 31, 2013.

31. David Yoel, Steven Walker, James Elwell, and Gilbert Moore, "The First Getaway Special: How It Was Done," *Space World*, May 1983, 9; Moore, "Remarks at the Funeral of NASA's Get Away Special Program"; Gordon Harris, "Shuttle to Carry Family Lizard," *Today* (FL), March 16, 1978, 10A; L. R. Megill to Director of Financial Management, NASA, December 21, 1976, folder 8184, NHRC/HQ; R. Gilbert Moore to Charles H. Carter, April 18, 1977, with attachment, folder 19814, NHRC/HQ; 123 Cong. Rec. S7579–81 (daily ed. May 13, 1977).

32. MO/Director, STS Operations (Chester M. Lee) to SP/Manager, Shuttle Projects Office, Kennedy Space Center, "Technical Support for Getaway Special Payloads," October 28, 1977, folder 8183, NHRC/HQ.

33. See, for example, Chester M. Lee to Robert C. Macauley, March 9, 1978; and Chester M. Lee to Earl R. Nadeau, April 12, 1978; both (and dozens of similar ones) in folder 8184, NHRC/HQ.

34. Park Seed, "Seeds in Space," Park Seed, accessed April 30, 2023, http://parkseed.com/seeds-in-space/a/13/; NASA, "STS-6 Pre-Flight Get Away Special Briefing," January 1983, 8, box 4, STS-6 Mission Documents Subseries, Shuttle Series, JSCHC/UHCL.

35. NASA, "Japanese Newspaper to Sponsor Space Shuttle Experiment," news release 81-191, December 8, 1981, folder 8183, NHRC/HQ. The experiment was reflown on STS-8 because water in the experiment chamber froze prematurely on STS-6.

36. John Dingman, "Shuttle to Make Eye-in-Sky Movie," *Birmingham News*, August 20, 1982; Ladwig, interview by the author, January 31, 2013; Linda Billings, "Get Away

Special," *Air & Space*, March 1988, 48–51; NASA, *Get Away Special . . . The First Ten Years*, 13; NASA, "Space Shuttle to Carry Explorer Scout Experiments," news release 82-68, April 22, 1982, folder 8183, NHRC/HQ; Flores, "Here's a GAS of a Cargo Flight."

37. M/Associate Administrator for Space Flight (John F. Yardley) to 100/Director, Goddard Space Flight Center, "Support for Self-Contained Payload Users," May 3, 1977, folder 8184, NHRC/HQ; NASA, Goddard Space Flight Center, *Getaway Special (GAS) Small Self-Contained Payloads: Experimenter Handbook* (Greenbelt, MD: NASA, 1978), folder 8182, NHRC/HQ; Clarke Prouty, "Get Away Special: The Low-Cost Route to Orbit," n.d., 5, folder 8183, NHRC/HQ; Space Transportation System; Use of Small Self-Contained Payloads, 45 Fed. Reg. 73022 (November 4, 1980) (amending 14 C.F.R. pt. 1214); Chester M. Lee, "Meeting with Mr. Yardley on the Small Self-Contained Payload Priorities and Amplification of Policy, March 20, 1978," memorandum, March 28, 1978, folder 8184, NHRC/HQ.

38. S-1/Associate Administrator for Space Science (Noel W. Hinners) to LC-5/Legislative Affairs Officer, "Student Involvement in NASA Programs," April 20, 1978 [date handwritten], folder 19815, NHRC/HQ; MO-6/Director, STS Operations (Chester M. Lee) to MT-3/Director, "Advanced Programs, Your Memorandum Dated April 24, 1978, Same Subject," May 1978 [day illegible], NHRC/HQ; MO/Director, STS Operations (Chester M. Lee) to PA/Acting Manager, Shuttle Payload Integration and Development Program Office, Johnson Space Center, "Accommodations for Small Payloads," June 10, 1977, folder 8184, NHRC/HQ.

39. Chester M. Lee to Patti Mancini, October 20, 1977, folder 8184, NHRC/HQ. Lee does not specify the source of opposition within NASA, but it almost certainly arose from Johnson Space Center, which was concerned about mission safety and interfering with the astronauts' time for other duties. Ladwig, interview by the author, January 31, 2013.

40. Robert Horvitz, "Art into Space: Artists Using Space Technology to Launch Aesthetics into Orbit," *Whole Earth Review*, Fall 1985, 27–28; Ladwig, interview by the author, January 31, 2013; Abrahamson, "Can the Space Shuttle Compete?"; NASA, "Satellites to Be Deployed from GAS Containers," news release 85-35, March 7, 1985, folder 8183, NHRC/HQ; NASA Daily Activities Report, May 8, 1985, folder 8183, NHRC/HQ; Prouty, "Get Away Special," 6.

41. Prouty, "Get Away Special," 6.

42. NASA, *Skylab: Classroom in Space*, ed. Lee B. Summerlin, NASA SP-401 (Washington, DC: US Government Printing Office, 1977), accessed April 30, 2023, http://history. nasa.gov/SP-401/contents.htm; NASA George C. Marshall Space Flight Center, *Skylab Student Project Summary Description*, MSFC-SL-73-3 (Marshall Space Flight Center, AL: NASA, 1973), 1, box 30, Robert Parker Files Subseries, Center Series, JSCHC/UHCL; M/Associate Administrator for Manned Space Flight (Dale D. Myers) to A/Administrator, "High School Student Participation in the Skylab Missions," August 18, 1971, folder 19816, NHRC/HQ.

43. M/Associate Administrator for Manned Space Flight to A/Administrator, "High School Student Participation in the Skylab Missions;" AD/Deputy Administrator (George M. Low) to F/Assistant Administrator for Public Affairs, "Use of NASA Gift Fund for Skylab Student Experiments," April 24, 1973, folder 5851, NHRC/HQ; Glen P. Wilson, "Outline of a Shuttle Student Involvement Program," April 1979, 8–10, folder 19815, NHRC/HQ.

44. LF-13/Special Assistant (Education) for the Director of Public Affairs (Frederick B. Tuttle), "Space Shuttle Student Project and Its Education-Related Activities," memorandum for the record, February 8, 1979, folder 19814, NHRC/HQ; H. A. Engle and D. L. Christensen, *Educational Planning for Utilization of Space Shuttle (ED-PLUSS)*, Executive Summary: Identification and Evaluation of Educational Uses and Users for the STS (Huntsville: University of Alabama in Huntsville, February 1975), 10, folder 19814, NHRC/HQ; Robert Staehle, "STS Operational Era Student Experiment Program," presentation charts, June 1976, folder 19813, NHRC/HQ; PA01 (James T. Murphy), Marshall Space Flight Center, to Philip Culbertson, NASA Headquarters, "Student Experiments for Shuttle OFT Flights and STS Operational Era," June 29, 1976, folder 19815, NHRC/HQ; O/Special Assistant to the Assistant Administrator for Planning and Program Integration (Robert G. Wilson) to P/Director, University Affairs, "Student Experiments on STS," July 15, 1976, folder 19815, NHRC/HQ.

45. Roger D. Kaul to Robert S. Tiemann, June 10, 1975, folder 19814, NHRC/HQ; Martin Marietta, "An Education Program for the Shuttle Era," presentation charts, September 21, 1976 [handwritten], folder 19814, NHRC/HQ; Forum for the Advancement of Students in Science and Technology, "FASST Workshop Spurs Planning for Student Competition," news release, December 18, 1979, folder 19813, NHRC/HQ; Federation of Americans Supporting Science and Technology, "A Proposal for a National Clearinghouse for Student Space Shuttle Payload Ideas," n.d., 5–6, folder 19814, NHRC/HQ.

46. William A. Huff to David Fradin, January 3, 1974; Frank E. Moss to James C. Fletcher, August 9, 1976; James C. Fletcher to Frank E. Moss, September 2, 1976; all in folder 19814, NHRC/HQ.

47. L/Associate Administrator for External Affairs (Herbert J. Rowe) to A/Administrator, "Space Shuttle Student Project," February 7, 1977, folder 19815, NHRC/HQ.

48. Leonard David to Robert Frosch, July 8, 1977; J. Jeffrey Irons to Robert A. Frosch, July 13, 1977; Dion W. J. Shea to Robert A. Frosch, August 1, 1977; all in folder 19813, NHRC/HQ.

49. Adlai E. Stevenson and Harrison Schmitt to Robert A. Frosch, December 14, 1977, folder 19812, NHRC/HQ; Comm. on Commerce, Science, and Transportation, NASA Authorization for Fiscal Year 1979, S. Rep. 95-799, at 59 (1978).

50. Robert A. Frosch to Harrison H. Schmitt, April 26, 1979, folder 19814, NHRC/HQ; Robert A. Frosch, statement, November 30, 1979, Appendix L in FASST, "Final Report: Workshop to Develop a National College-Level Student Experiment Competition for the Space Shuttle," January 31, 1980, folder 19812, NHRC/HQ.

51. FASST, "An Unsolicited Proposal to Support a Workshop to Develop a National College-Level Student Experiment Contest for the Space Shuttle," September 1979; Glen P. Wilson to Alan Ladwig, September 28, 1979; FASST, "Final Report: Workshop to Develop a National College-Level Student Experiment Competition for the Space Shuttle," January 31, 1980; all in folder 19812, NHRC/HQ.

52. NASA, "Selection Begun in Shuttle Student Project Competition," news release 81-31, February 24, 1981, folder 19816, NHRC/HQ; Student Activities Steering Committee, summary minutes, August 28, 1979, 4, folder 19812, NHRC/HQ; Charles Pellerin to Mr. Culbertson, "Student Program," draft memo, June 21, 1977, folder 19815, NHRC/HQ; SB-3/ Acting Deputy Director for Life Sciences (Richard S. Young) to LF-13/NASA Consultant, "Spacelab Accommodations for Student Experiments' Draft Letter—Yardley to Lovelace," June 6, 1979, folder 19816, NHRC/HQ.

53. Student Activities Steering Committee, summary minutes; L/Special Assistant for Student Activities (Glen P. Wilson) to L/Acting Associate Administrator for External Relations, "Selection of Contractor for SSIP-S," May 1, 1980, folder 19815, NHRC/HQ; Glynn S. Lunney to ADB/Bob Allnut, June 20, 1980, folder 19815, NHRC/HQ.

54. NASA considered foreign participants but decided to restrict eligibility to US citizens because of issues such as travel costs and language barriers. See Wilson, "Outline of a Shuttle Student Involvement Program," 14.

55. NASA, "Student's Experiment to Fly on Third Shuttle Mission," news release 82-27, February 24, 1982, folder 19815, NHRC/HQ; NASA, "Shuttle Student Project Winners Meet with NASA Scientists," news release 81-125, August 20, 1981, folder 19816, NHRC/ HQ.

56. NASA, "Student's Experiment to Fly on Third Shuttle Mission," press kit, February 1982, folder 19815, NHRC/HQ; Alan Ladwig to Jack Loosbrock, September 17, 1981, folder 19816, NHRC/HQ; M/Associate Administrator for Space Transportation Systems (James A. Abrahamson) to JSC/Manager, Space Transportation Systems Operations, "Shuttle Student Involvement Project for Secondary Schools (SSIP-S)," January 22, 1982, folder 19815, NHRC/HQ; Shuttle Student Involvement Program Task Force, "Space Shuttle Student Involvement Program (SSIP) Task Force Analysis," February 8, 1985, folder 19816, NHRC/HQ; "Space Science Student Involvement Program, 1980–1990: Where Are They Now?" 1990, folder 19816, NHRC/HQ.

57. MP-5/SSIP Program Manager (Alan Ladwig) to M/Associate Administrator for Office of Space Transportation Systems, "A. JSC Project Office for SSIP; B. Lunney Memo of May 11; C. Soffen Memo of May 21; D. Code M SSIP Responsibility," June 2, 1982; EB-3/ Director, Life Sciences Division (Gerald A. Soffen) to M/Associate Administrator for Space Transportation Systems and E/Associate Administrator for Space Science and Applications, "Student Experiments Involving Rats," May 21, 1982; both in folder 19815, NHRC/HQ.

58. M/Associate Administrator for Space Transportation Systems (James A. Abrahamson) to Johnson Space Center, Kennedy Space Center, and Marshall Space Flight Center, "Shuttle Student Involvement Program," August 9, 1982, folder 19815, NHRC/HQ; M/Associate Administrator for Space Transportation Systems to JSC/Manager, Space Transportation Systems Operations, "Shuttle Student Involvement Project for Secondary Schools (SSIP-S)"; James A. Abrahamson to Glynn S. Lunney, February 22, 1982, folder 19815, NHRC/HQ.

59. Reagan, "Address before a Joint Session of the Congress," 90.

60. Jim Alston, interview by the author, April 11, 2014.

61. NASA, *SEEDS: A Celebration of Science*, NASA EP-281 (Washington, DC: US Government Printing Office, 1991), v; NASA, "NASA Space Shuttle to Carry Tomato Seeds on Mission 41-C," news release 84-47, n.d., NHRC online, https://historydms.hq.nasa.gov/sites/default/files/DMS/e000041998.pdf.

62. Alston, interview by the author, April 11, 2014; Doris Grigsby, interview by the author, April 11, 2014; NASA, "NASA Space Shuttle to Carry Tomato Seeds on Mission 41-C."

63. Park Seed, "Seeds in Space"; NASA, *SEEDS*.

64. Thomas O'Toole, "Gerbils, Ashes Nixed in Shuttle Bids," *Today* (FL), January 21, 1978, A1, folder 8164, NHRC/HQ; Chester M. Lee to Udo Schlegel, August 25, 1978, folder 5310, NHRC/HQ; Lee to Albert Notarbartolo, December 19, 1978, folder 5310, NHRC/HQ; John F. Murphy to Senator Howell Heflin, May 2, 1984, folder 5308, NHRC/HQ.

65. "Moon Billboards," *Omni*, November 1981, 42.

66. See, for example, Thomas O'Toole, "Ex-Astronauts Disregarded Warning against 'Souvenirs,'" *Washington Post*, August 1, 1972, A8.

67. See, for example, Joan Johnson-Freese, "Launch Dazzling but Cut Out P.R., Commercialism," *Orlando Sentinel*, November 30, 1985; O'Toole, "Gerbils, Ashes Nixed in Shuttle Bids."

68. Articles Authorized to Be Carried on Space Transportation System Flights, 43 Fed. Reg. 49979 (October 26, 1978) (amending 14 C.F.R. pt. 1214). Later revisions designated the Associate Administrator for Space Flight as the final authority.

69. NASA, "Getaway Special," fact sheet, March 1981, folder 8183, NHRC/HQ. NASA made an exception to its rule prohibiting use of GAS cans to fly payloads for resale. In 1983 STS-8 carried nearly a quarter-million commemorative postal covers for the United States Postal Service to sell to collectors in honor of NASA's twenty-fifth anniversary for fifteen dollars apiece, with the revenue to be split between the two agencies. Room for the covers became available when a technical problem prevented the mission from carrying the satellite originally scheduled to fly in the shuttle's cargo bay.

70. "Astro Burial," *Omni*, February 1980, 39; Joseph M. Roberto to Donna S. Miller, August 6, 1979; Roberto to Chester M. Lee, August 20, 1979; Lee to Roberto, August 27, 1979; John F. Yardley to Roberto, October 15, 1979; all in folder 5382, NHRC/HQ.

71. Julian Scheer to Edward S. Stephenson, October 10, 1966, folder 1904, NHRC/HQ; Volkswagen, "Space Shuttle" (advertisement), *New York Times*, April 15, 1981; *Playboy*, "Confidence. America's Got It Again" (advertisement), *New York Times*, May 21, 1981; Kevin Anderson, "NASA Promotion Won't Fly without Approval of Agency," *USA Today*, December 6, 1983; Frank Johnson to Mike Weeks, handwritten note, October 22, 1985, folder 11903, NHRC/HQ; Frank S. Johnson Jr., to Charles S. Adorney, n.d., folder 1134, NHRC/HQ; James Fisher, "NASA Image Can Put Sales in Orbit," *Orlando Sentinel*, June 6, 1985.

72. Robert H. Lorsch to Jesse W. Moore, July 3, 1984, with attachment; "Moon Billboards," 42; Brian M. Duff to Bob Lorsch, November 25, 1980; both in folder 11903, NHRC/HQ.

73. Terence T. Finn to Don Fuqua, December 10, 1980; Finn to Keith Luse, March 11, 1981, folder 5382, NHRC/HQ; Yardley to Roberto, October 15, 1979; Lee to Roberto, August 27, 1979; "Astro Burial."

74. Smith, interview by the author, June 10, 2013; Horvitz, "Art into Space," 31.

75. AD/Deputy Administrator (A. M. Lovelace) to M/Associate Administrator for Space Transportation Systems and P/Chief Scientist, "Approval to Establish Committee to Review Art and Structure Proposals to Fly as Self-Contained Payloads on the Space Shuttle," n.d., folder 8183, NHRC/HQ; Smith, interview by the author, June 10, 2013; Ladwig, interview by the author, January 31, 2013.

76. Nonscientific Payloads, 49 Fed. Reg. 34445 (August 31, 1984) (amending 14 C.F.R. pt. 1214); Horvitz, "Art into Space," 30.

77. David Stipp, "Spaced-Out Artwork Adds Allure to Flights of NASA's Shuttles," *Wall Street Journal*, December 10, 1985; Charlie Jean, "Artist's Little World to Fly on Discovery," *Orlando Sentinel*, February 20, 1989; Horvitz, "Art into Space," 29–30; NASA, "Art Work to Be First Shuttle Non-scientific Payload," news release 85-175, December 20, 1985, folder 8163, NHRC/HQ; "Lowry Burgess: Artist, Educator," *Folly*, August 2009, 4–5.

Chapter 6: Creating Space for New Flyers

1. See, for example, Walter Cunningham, *The All-American Boys* (New York: Macmillan, 1977) and Roger D. Launius, "Heroes in a Vacuum: The Apollo Astronaut as Cultural Icon," *Florida Historical Quarterly*, 87, no.2 (2008): 174–209.

2. "Astronaut Selection and Training," NASA Facts, Lyndon B. Johnson Space Center, JSC-11863, September 1979, folder 8929, NASA Historical Reference Collection, NASA Headquarters, Washington, DC (hereafter cited as NHRC/HQ); Tom Wolfe, *The Right Stuff* (New York: Picador, 1979), 148.

3. Hugh L. Dryden to Evan H. Walker, October 16, 1962, folder 8947, NHRC/HQ; National Research Council, Space Science Board, *A Review of Space Research* (Washington, DC: National Academy Press, 1962); W. David Compton and Charles D. Benson, *Living and*

Working in Space: A History of Skylab, NASA SP-4208 (Washington, DC: US Government Printing Office, 1983), 65.

4. "Scientist Astronauts," March 13, 1964, folder 8954, NHRC/HQ; John W. Finney, "NASA to Recruit Scientists to Fly with Astronauts," *New York Times*, October 18, 1964, 1; NASA, "NASA Reports Some 900 Persons Interested in Scientist-Astronaut Program," news release 64-315, December 16, 1964, folder 8954, NHRC/HQ; NASA, "NASA Selects Six Scientists-Astronauts for Apollo Program," news release 65-212, June 28, 1965, folder 8954, NHRC/HQ; NASA, "Eleven Civilian Scientist-Astronauts Join NASA Team," news release 67-211, August 4, 1967, folder 8956, NHRC/HQ.

5. Joseph P. Allen, interview by Jennifer Ross-Nazzal, Washington, DC, March 16, 2004, transcript, Johnson Space Center Oral History Project, Houston, TX, https://historycollection.jsc.nasa.gov/JSCHistoryPortal/history/oral_histories/AllenJP/AllenJP_3-16-04.htm.

6. Homer E. Newell, "Meeting at Houston between Headquarters and Houston Management and the Scientist-Astronauts on April 8, 1971," April 19, 1971, folder 8945, NHRC/HQ.

7. "Report on the Subcommittee on Scientists Astronauts of the NASA Space Program Advisory Council," 1975, referenced in Matthew H. Hersch, *Inventing the American Astronaut* (New York: Palgrave Macmillan), 2012, 148; Frederick Seitz, "Letter to George M. Low," September 8, 1975, folder 4159, NHRC/HQ; Nathaniel B. Cohen to distribution, "Transmittal of Minutes of January 15, 1975, SPAC Ad Hoc Subcommittee on Scientist Astronauts Meeting," 4, with attachment, "Summary Minutes of the SPAC Ad Hoc Subcommittee on Scientist Astronauts, January 15, 1975," 2, April 3, 1975, folder 8948, NHRC/HQ; John E. Naugle to Charles Berry, June 27, 1975, folder 8945, NHRC/HQ; NASA, *Opportunities as Candidates for Mission Specialist Astronauts*, n.d., folder 8946, NHRC/HQ; Payload Specialists for NASA or NASA-Related Payloads, 43 Fed. Reg. 9790 (March 10, 1978) (amending 14 C.F.R. pt. 1214).

8. Christopher C. Kraft Jr., to John E. Naugle, September 3, 1976, folder 8959, NHRC/HQ; NASA, "Requirements May Ease for Shuttle Non-pilot Crew Members," news release 75-79, March 24, 1975, NHRC/HQ.

9. NASA Astronaut Candidate Recruitment and Selection Program, 44 Fed. Reg. 36024 (June 20, 1979) (amending 14 C.F.R. pt. 1214); NASA, "NASA Selects 35 Astronaut Candidates," news release 78-7, folder 8945, NHRC/HQ; Kraft to Naugle, September 3, 1976; NASA, *Opportunities as Candidates for Mission Specialist Astronauts*; NASA/Lyndon B. Johnson Space Center, *Announcement No. 3ACS-83 for Mission Specialist and Pilot Astronaut Candidates*, 1983, folder 19758, NHRC/HQ.

10. See Margaret A. Weitekamp, *Right Stuff, Wrong Sex: America's First Women in Space Program* (Baltimore, MD: Johns Hopkins University Press, 2004) and Bettyann Holtzmann Kevles, *Almost Heaven: The Story of Women in Space* (Cambridge: MIT Press, 2006).

11. Brian Duff, interview by John Mauer, May 24, 1989, transcript, box 12, folder 16, Glennan-Webb-Seamans Project, National Air and Space Museum Archives, Smithsonian Institution, Washington, DC; Frank Macomber, "With Man on Moon, Can Space Lady Be Far Behind?," *San Jose Mercury*, July 26, 1972; "Women Aid Astronauts, Not Yet Needed in Spacecraft," *Telegram* (Bridgeport, CT), June 26, 1963; "Women May Share Man's Space Trips," *Huntsville News*, February 10, 1970, folder 8995, NHRC/HQ.

12. H. E. Van Ness, "Status of Negro Applicants for Astronaut Selection," July 9, 1962, folder 8993, NHRC/HQ. According to Van Ness, NASA's assistant director for manned space flight operations, an army test pilot qualified but was not recommended by the army due to "disciplinary problems." Meanwhile, US Air Force Captain Edward Dwight, a Black man, was recommended by the air force but NASA passed over him. J. K. Obatala, "We Need to Correct a Space Age Injustice: America Can Still Have a Black Astronaut," *Los Angeles Times*, March 21, 1974.

13. Simpson to Duge, October 11, 1962, folder 8993, NHRC/HQ; Julian Scheer to Virginia Allan, August 28, 1963, folder 8996, NHRC/HQ.

14. Hugh L. Dryden to John W. Wydler, July 1, 1963, folder 8996, NHRC/HQ.

15. Joseph Morgenstern, "What's It to Us?," *Newsweek*, July 7, 1969, 64, 67, 68; Obatala, "We Need to Correct a Space Age Injustice"; "A Plea for US Spacewomen," *Boston Globe*, February 23, 1971.

16. Ruth Bates Harris to Robert J. Brown, May 30, 1972, folder 8993, NHRC/HQ; "NASA Defends Lag on Women's Jobs," *New York Times*, January 12, 1974.

17. Kim McQuaid, "Race, Gender, and Space Exploration: A Chapter in the Social History of the Space Age," *Journal of American Studies* 41, no. 2 (2007): 420; NASA, *Wings in Orbit: Scientific and Engineering Legacies of the Space Shuttle*, ed. Wayne Hale, Helen Lane, Gail Chapline, and Kamlesh Lula, NASA SP-2010-3409 (Washington, DC: US Government Printing Office, 2010), 462.

18. NASA, "NASA Administrator Cites Equal Employment Advances," news release 74-133, folder 4215, NHRC/HQ; James C. Fletcher to Earle F. Kyle Jr., November 23, 1971, folder 8993, NHRC/HQ; Ruth Bates Harris, March 7, 1972, with attachment, "Remarks by NASA Administrator Dr. James C. Fletcher at the Equal Employment Opportunity Conference, Kennedy Space Center, March 2, 1972," 5, NHRC online, accessed May 1, 2023, https://historydms.hq.nasa.gov/sites/default/files/DMS/e000038126%20-%20508%20pass.pdf.

19. Dudley G. McConnell to Vernon A. Burford, April 22, 1974, folder 8996, NHRC/HQ.

20. Dudley G. McConnell to Gemma Arnott, April 19, 1974, folder 8996; NASA, "Bathroom Commode Design for Space Shuttle Passengers," news release 72-163, August 9, 1972, folder 8994; NASA, "Women as Shuttle Passengers under Study," news release 73-185, September 17, 1973, folder 8994; NASA, "Bay Area Women in Space Tests," news release 77-15, March 24, 1977, folder 8994; NASA, "Women to Begin Tests to Assess Fitness for Space Travel," news release 77-74, April 7, 1977, folder 8994; all NHRC/HQ.

21. Even so, uses of male-oriented language lingered for some time. For example, a 1974 NASA overview of the shuttle uses the term "crewmen" and contains a sketch of four men in the orbiter's flight deck. See NASA, *Space Shuttle Program Overview* (Washington, DC: US Government Printing Office, 1974), 10, box 4, STS General Information Subseries, Shuttle Series, Johnson Space Center History Collection, University of Houston–Clear Lake, Houston, TX (hereafter cited as JSCHC/UHCL). NASA eventually stopped using the term "manned" when referring to the human space program.

22. John P. Donnelly to A/Administrator (Fletcher), "Black Astronaut," April 24, 1974, folder 8993, NHRC/HQ. Donnelly may have recalled that US Air Force Major Robert Lawrence, who was Black, had been selected as part of the air force's Manned Orbiting Laboratory (MOL) astronaut corps. Had Lawrence not been killed in a 1967 flying accident, he likely would have been transferred to NASA in 1969, as were the other MOL astronauts when the Air Force cancelled the MOL program. See Donnelly to Elizabeth A. Cooper, July 13, 1972, folder 8993, NHRC/HQ.

23. Henry E. Clements to AA/Associate Administrator (Petrone) and AAD/Deputy Associate Administrator, "Black and Female Astronauts," folder 8993, NHRC/HQ; NASA, "NASA to Recruit Space Shuttle Astronauts," news release 76-122, July 8, 1976, folder 8945, NHRC/HQ; T. A. Heppenheimer, *History of the Space Shuttle*, vol. 2, *Development of the Shuttle, 1972–1981* (Washington, DC: Smithsonian Institution Press, 2002), 389.

24. George M. Low to M/Associate Administrator for Space Flight (Yardley), "Astronaut Selection Program," December 15, 1975, folder 8993, NHRC/HQ; James C. Fletcher to Tuskegee Airmen, Inc., Attending Their Annual Convention in Philadelphia, August 18, 1976, folder 8993, NHRC/HQ; QP-1/Manager, Equal Opportunity Professional Recruiting Office (Alfred Clinkscales) to QP/Director of Personnel, "Project Status Report for QP-1," January 5, 1977, folder 8959, NHRC/HQ; Heppenheimer, *History of the Space Shuttle*, 389.

25. QP-1/Manager, Equal Opportunity Professional Recruiting Office (Alfred Clinkscales) to distribution: regional coordinators, "Recruitment for Space Shuttle Astronauts," November 23, 1976; QP-1/Manager, Equal Opportunity Professional Recruiting Office (Alfred Clinkscales) to QP/Director of Personnel, "Activity Report for QP-1," March 11, 1977; QP-1/Manager, Equal Opportunity Professional Recruiting Office (Alfred Clinkscales) to QP/Director of Personnel, "Status Report for QP-1," March 28, 1977; all in folder 8959, NHRC/HQ.

26. Heppenheimer, *History of the Space Shuttle*, 389; Joseph P. Allen to Alan Cranston, July 12, 1978, folder 8945, NHRC/HQ.

27. NASA, "Television Celebrity to Help NASA Recruit Shuttle Astronauts," news release 77-32, February 24, 1977; Women in Motion, Inc., Final Report for NASA, August 10, 1977, 12–13; both in folder 8935, NHRC/HQ.

28. "#2 NASA Public Service Announcement, N. Nichols," with cover memo, June 24, 1977, folder 8935, NHRC/HQ.

29. Robert Churchwell, "Star Trek's Uhura Fights Racism in NASA," *Nashville Banner*, March 18, 1977; "Sample Presentation, Professional Organization," n.d.; both in folder 8935, NHRC/HQ.

30. NASA, "NASA Selects 35 Astronaut Candidates," news release 78-7, January 16, 1978, folder 8945, NHRC/HQ; QP-1/Manager, Equal Opportunity Professional Recruiting Office (Alfred Clinkscales) to QP/Director of Personnel, "The Shuttle Program," June 15, 1977, folder 8959, NHRC/HQ; Allen to Cranston, July 12, 1978; Heppenheimer, *History of the Space Shuttle*, 390.

31. Earl Lane, "Skylab Promises Humanized Astronauts," *Los Angeles Times*, November 12, 1978; Jim Maloney, "Shuttle Precedents Secondary," *Houston Post*, April 30, 1982, 17A, folder 8958, NHRC/HQ; Associated Press, "Shuttle—Black Reaction," August 30, 1983, folder 8993, NHRC/HQ.

32. NASA, "2,937 Apply for Astronaut Positions," news release 79-175, December 13, 1979, folder 8959, NHRC/HQ; NASA, "NASA Selects 19 Astronaut Candidates," news release 80-78, May 29, 1980, folder 8945, NHRC/HQ; JoAnn H. Morgan and James M. Ragusa, "Women: A Key Work Force in Preparation for Space Flight," June 26, 1972, exhibit 3, folder 8995, NHRC/HQ; NASA, "NASA Features Woman Commentator for Shuttle Flights," news release 77-84, April 25, 1977, NHRC online, https://historydms.hq.nasa.gov/sites/default/files/DMS/e000021075.pdf.

33. Al Stewart, letter to the editor, *Aviation Week & Space Technology*, September 9, 1985, 120; Philip M. Boffey, "Despite Numbers of Applicants, Few Civilians Are Selected as Astronauts," *New York Times*, August 17, 1987, A8.

34. Kevin Shyne, "High School Grads Can Train to Fly Space Shuttle," *USA Today*, October 6, 1983.

35. James C. Fletcher, "Coming—The Space Shuttle," *Saturday Evening Post*, May/June 1973, 68; National Research Council, *A Review of Space Research*; Hans Mark to Nathaniel B. Cohen, February 21, 1975, folder 8948, NHRC/HQ.

36. AA/Associate Administrator (John E. Naugle) to M/Associate Administrator for Space Flight, "Mission Specialist Selection Process," December 30, 1976, with attachment, "Space Shuttle Crew Policy Issues," folder 8945, NHRC/HQ; Christopher C. Kraft Jr. to James C. Fletcher, October 12, 1976, folder 8945, NHRC/HQ; Kraft to Alan M. Lovelace, May 10, 1978, folder 8946, NHRC/HQ.

37. Payload Specialists for NASA or NASA-Related Payloads, 43 Fed. Reg. 9790 (March 10, 1978) (amending 14 C.F.R. pt. 1214); NASA, *The Role of the Payload Specialist in the Space Transportation System for NASA or NASA-Related Payloads* (n.d.), folder 8971, NHRC/HQ.

38. Kraft to Lovelace, May 10, 1978.

39. "NASA Restudies Process for Picking Shuttle Crew," *Aviation Week & Space Technology*, June 19, 1978, 31; Payload Specialists for NASA or NASA-Related Payloads, 45 Fed. Reg. 8001 (February 6, 1980) (amending 14 C.F.R. pt. 1214).

40. Jon M. Smith, interview by the author, June 10, 2013; John F. Yardley, "Office of Space Transportation Systems Status Summary," statement for the record to the Subcommittee on Space Science and Applications, Committee on Science and Technology, US House of Representatives, September 25, 1978, NHRC/HQ; Alex P. Nagy, "Passenger Flight on Space Shuttle" (draft), May 19, 1978, attachment to Alex P. Nagy to John Naugle, December 7, 1982, folder 8074, NHRC/HQ.

41. James M. Beggs to Harrison H. Schmitt, October 7, 1982, folder 8929, NHRC/HQ. A handwritten note on the letter indicates that NASA sent identical letters to its other oversight committees.

42. Charles D. Walker, interview by Jennifer Ross-Nazzal, Washington, DC, November 19, 2004, transcript, Johnson Space Center Oral History Project, Houston TX, https://historycollection.jsc.nasa.gov/JSCHistoryPortal/history/oral_histories/WalkerCD/WalkerCD_11-19-04.htm.

43. James A. Abrahamson, "Statement of Lieutenant General James A. Abrahamson, Associate Administrator for Space Flight, National Aeronautics and Space Administration, before the Subcommittee on Space Science and Applications, Committee on Science and Technology, House of Representatives," March 1, 1983, NHRC online, accessed May 1, 2023, https://historydms.hq.nasa.gov/sites/default/files/DMS/e000043019.pdf; Walker, interview by Ross-Nazzal, November 19, 2004; NASA, "Arabsat Payload Specialist Activities," news release 85-69, May 4, 1985, folder 8971, NHRC/HQ.

44. Robert C. Cowen, "Opening the Hatch to Foreign Astronauts," *Christian Science Monitor*, June 24, 1985, 8.

45. James Fisher, "Nelson Awaits Slot to Go on Shuttle Mission," *Orlando Sentinel*, August 8, 1985; Beggs, interview by the author, April 8, 2013.

46. John F. Murphy to Lawton Chiles, March 7, 1985, folder 10990, NHRC/HQ.

47. Chris Dubbs and Emeline Paat-Dahlstrom, *Realizing Tomorrow: The Path to Private Spaceflight* (Lincoln: University of Nebraska Press, 2011), 80–81.

48. Edward Crosby, "Why 'Astro-naughts' on Shuttle Flights?" *Florida Today*, September 15, 1985; Douglas Pike, "Nelson's Joy Ride into Space a Stunt We Don't Need," *Orlando Sentinel*, October 27, 1985; "Launching 'Buck Rogers' Nelson," *St. Petersburg Times*, September 11, 1985; "No Amateurs, Please," *Augusta (GA) Herald*, September 18, 1985.

49. Henry W. "Hank" Hartsfield Jr., interview by Carol Butler, Houston, TX, June 12, 2001, transcript, Johnson Space Center Oral History Project, Houston, TX, accessed May 1, 2023, https://historycollection.jsc.nasa.gov/JSCHistoryPortal/history/oral_histories/HartsfieldHW/HartsfieldHW_6-12-01.htm.

50. Charles D. Walker, interview by Sandra Johnson, Houston, TX, April 14, 2005, transcript, Johnson Space Center Oral History Project, Houston, TX, accessed May 1, 2023, https://historycollection.jsc.nasa.gov/JSCHistoryPortal/history/oral_histories/WalkerCD/WalkerCD_4-14-05.htm; Walker, interview by Ross-Nazzal, November 19, 2004;

Walker, interview by Sandra Johnson, Springfield, VA, November 7, 2006, transcript, Johnson Space Center Oral History Project, Houston TX, https://historycollection.jsc.nasa.gov/JSCHistoryPortal/history/oral_histories/WalkerCD/WalkerCD_11-7-06.pdf.

51. Walker, interview by Johnson, November 7, 2006.

52. Steven Mufson, "Pan Am Still Plans Moon Flight, and 93,000 Are on Waiting List," *Washington Post*, July 22, 1989; George M. Low to Hugh Downs, April 5, 1971, folder 498, NHRC/HQ.

53. Amy Paige Kaminski, "Explorers We? The Making, Unmaking, and Public Involvement Legacy of NASA's Space Flight Participant Program," *Quest: The History of Space Flight Quarterly* 20, no. 1 (2013): 12–21; Chet Jezierski to James Beggs, July 22, 1985, folder 19758, NHRC/HQ; Ken Jopp, letter to the editor, *Aviation Week & Space Technology*, November 4, 1985, 96.

54. M-N/Public Information Officer for Space Flight (David W. Garrett) to F/Acting Assistant Administrator for Public Affairs (Robert J. Shafer), "Space Shuttle 'Unique Personality,'" June 28, 1976, folder 8073, NHRC/HQ. The memo cites the "Aerospace Advisory Board (ASEB)," but no group by this exact name is known to have existed at NASA. Given the acronym, it is most likely a reference to the Aeronautics and Space Engineering Board of the National Research Council.

55. Kenneth F. Weaver to James C. Fletcher, March 8, 1976, folder 8078, NHRC/HQ ; John L. Hammersmith, "Note for Bill Land," n.d., with attachment, Martin Company, "The National Geographic Society Space Expedition," May 26, 1965, folder 19758, NHRC/HQ; Fletcher to Weaver, March 24, 1976, folder 8078, NHRC/HQ; Fletcher to F (Deputy Assistant Administrator for Public Affairs), March 16, 1976, folder 8078, NHRC/HQ.

56. "Non-astronaut Flight Participation in OFT: Report of Ad Hoc Study Team," July 1976, box 4, STS General Information Subseries, Shuttle Series, JSCHC/UHCL; John L. Hammersmith to General McNickle, "Non-crew Flight Participation: Selection Process," February 15, 1977, folder 19758, NHRC/HQ; Nagy, "Passenger Flight on Space Shuttle" (draft); Arnold W. Frutkin to LF-6/Robert Newman, "Draft NMI on Shuttle Passengers," June 7, 1978, attachment to Alex P. Nagy to John Naugle, December 7, 1982, folder 8074, NHRC/HQ.

57. Richard C. Levy, "Everybody Is Lining Up for Space," *Parade*, December 30, 1979, 4–5; Tom DeVries, "A Shot at Space," *Sun* (Mississippi Gulf Coast), April 16, 1981; "The Plane That Will Take Us Out of This World," *American Legion Magazine*, February 1977, 7; LF-6/Director of Public Affairs (Robert A. Newman) to L/(Acting) Associate Administrator for External Relations (Arnold W. Frutkin), "Draft NMI on Shuttle Passengers," November 6, 1978, attachment to Alex P. Nagy to John Naugle, December 7, 1982, folder 8074, NHRC/HQ.

58. Beggs, interview by the author, April 8, 2013; James M. Beggs, "Should NASA Fly Private Citizens aboard the Space Shuttle?" *NASA Activities*, October 1982, 3, folder

19754, NHRC/HQ; NASA Advisory Council Informal Task Force for the Study of Issues in Selecting Private Citizens for Space Shuttle Flight, "Summary Minutes from July 22, 1982, Meeting," folder 16713, NHRC/HQ; Carl R. Praktish to Sylvia D. Fries, September 28, 1982, folder 16713, NHRC/HQ.

59. Florence R. Skelly to John E. Naugle, September 2, 1982, folder 16713, NHRC/HQ; "Sectors of the NASA Public—Hypotheses on Shuttle Passengers," n.d., folder 16713, NHRC/HQ; Richard Gilluly, "NASA Lining Up the Shuttle Bugs," *Los Angeles Times*, November 18, 1982.

60. John E. Naugle to Jake Garn, January 13, 1983, folder 10990, NHRC/HQ; "Report of the Informal Task Force for the Study of Issues in Selecting Private Citizens for Space Shuttle Flight," June 16, 1983, folder 19759, NHRC/HQ.

61. Jim Detjen, "NASA to Choose Citizen for Shuttle," *Boston Sunday Globe*, September 18, 1983, folder 8072, NHRC/HQ.

62. Citizen Observers/Participants, 48 Fed. Reg. 56770 (December 23, 1983) (amending 14 C.F.R. pt. 1214); Alan Ladwig, interview by the author, Washington, DC, December 28, 2010; Frank Hyland, "Get in Line Now to Hitch a Ride on Space Shuttle," *Atlanta Journal*, August 23, 1979; "NASA Curbing Space-Flight Merchandising," *New York Times*, May 13, 1984.

63. Citizen Observers/Participants, 48 Fed. Reg. 56770. Copies of the twenty-two public responses are in folder 8074, NHRC/HQ.

64. Dubbs and Paat-Dahlstrom, *Realizing Tomorrow*, 78; Space Flight Participants, 49 Fed. Reg. 17736 (April 25, 1984) (amending 14 C.F.R. pt. 1214).

65. Thomas O'Toole, "Reagan Tells NASA to Choose Schoolteacher for Shuttle Flight," *Washington Post*, August 28, 1984, A1; M/Associate Administrator for Space Flight (James A. Abrahamson) to LB/Executive Secretary, NASA Advisory Council, "First Meeting of the Citizen Observer/Participant (COP) Evaluation Committee" (draft), 1984 [handwritten in corner], folder 19759, NHRC/HQ.

66. Arlen J. Large, "'Dear NASA,' They Write, 'Won't You Take Me Along?'" *Wall Street Journal*, April 25, 1983; Michael Kernan, "Shuttle: Everyone Wants to Ride," *Los Angeles Times*, December 11, 1983. Folder 19758 in the NHRC/HQ contains many public letters expressing interest in flying aboard the shuttle. Ladwig, interview by the author, December 28, 2010.

67. National Commission on Excellence in Education, *A Nation at Risk: The Imperative for Educational Reform* (Washington, DC: US Government Printing Office, 1983); NASA, "New NASA Educational Program for Science and Math Announced," news release 84-83, June 19, 1984, NHRC online, https://historydms.hq.nasa.gov/sites/default/files/DMS/e000041999.pdf; ADB/Associate Deputy Administrator (Philip E. Culbertson) to A/Administrator (James M. Beggs), "Proposal for First Flights of Private Citizens," April 4, 1984, folder 19758, NHRC/HQ; James M. Beggs, interview by Kevin M. Rusnak, Bethesda,

MD, March 7, 2002, transcript, NASA Oral History, NHRC, Washington, DC, https://historycollection.jsc.nasa.gov/JSCHistoryPortal/history/oral_histories/NASA_HQ/Administrators/BeggsJM/BeggsJM_3-7-02.htm; O'Toole, "Reagan Tells NASA to Choose Schoolteacher."

68. NASA/Council of Chief State School Officers, *Announcement of Opportunity: The NASA Teacher in Space Project* (Washington, DC, 1984), box 23, STS 51-L General History Files Subseries, General Reference Series, JSCHC/UHCL.

69. Alan Ladwig to Ms. Brown, March 13, 1985, folder 19758, NHRC/HQ.

70. NASA, *Teacher in Space Project* (Washington, DC: US Government Printing Office, 1985), box 1, STS 51-L Documents Subseries, Shuttle Series, JSCHC/UHCL; NASA, "Live Lessons, Mission Watch Highlight Shuttle Mission 51-L," news release 85-156, November 26, 1985, folder 19758, NHRC/HQ.

71. Patricia Palazzolo, interview by the author, Pittsburgh, PA, March 25, 2013; "Appendix B: Teacher in Space Program, Educational Activities of the Space Ambassadors, June 1985 through July 1986," n.d., folder 20198, NHRC/HQ.

72. NASA, "Journalist to Fly on Space Shuttle," news release 85-147, October 24, 1985, folder 19754, NHRC/HQ; NASA/Association of Schools of Journalism and Mass Communication, Journalist-in-Space Project application packet, 1985, folder 19753, NHRC/HQ.

73. O'Toole, "Reagan Tells NASA to Choose Schoolteacher"; Ladwig, interview by the author, December 28, 2010; Michael Mecham, "Disabled May Fly on Shuttle," *USA Today*, March 22, 1985; National Aeronautics and Space Administration Authorization Act of 1986, Pub. L. No. 99-170, 99 Stat. 1012 (1985).

74. See, for example, John F. Murphy to Lee B. Hamilton, May 10, 1985, and Murphy to Mark C. Munson, December 3, 1985; both in folder 19758, NHRC/HQ.

75. Ann Bradley to members of the Space Flight Participation Evolution Committee, August 24, 1984, with attachments, folder 19759, NHRC/HQ; Clifford Cunningham, "Civilian Space Travel is Nearing Reality," publication name illegible, February 22 [handwritten] [1984?], folder 8072, NHRC/HQ; Ladwig, interview by the author, January 31, 2013; Kevin Anderson, "2,000 Sign Up for Space Trip," *USA Today*, June 19, 1985; Leonard David, "Commuting into the Cosmos," *Space World*, August 1985, 11, 14.

Chapter 7: Reevaluating the Democratic Imaginary

1. Mark Chartrand and Leonard David, "James Beggs, NASA Administrator," *Insight*, November/December 1981, 6–7; Louis Harris, "Space Shuttle Considered Major Breakthrough for US Technology," Harris Survey news release, June 1, 1981, "Harris" folder, box 16, Alan Ladwig unprocessed papers, NASA Historical Reference Collection, NASA Headquarters, Washington, DC (hereafter cited as NHRC/HQ); Roger D. Launius, "Public Opinion Polls and Perceptions of US Human Space Flight," *Space Policy* 19, no. 3 (2003): 168.

2. Michael A. G. Michaud, *Reaching for the High Frontier: The American Pro-space Movement, 1972–84* (New York: Praeger, 1986), 294.

3. Alex Roland, "The Shuttle: Triumph or Turkey?" *Discover*, November 1985, 29–49.

4. Associated Press, "NASA Releases Shuttle Launch Schedule," *Greensboro (NC) News & Record*, April 7, 1985, A19, folder 8163, NHRC/HQ.

5. Diane Vaughan, *The Challenger Launch Decision: Risky Technology, Culture, and Deviance at NASA* (Chicago: University of Chicago Press, 1996).

6. Rather quoted in David Ignatius, "Did the Media Goad NASA into the Challenger Disaster?" *Washington Post*, March 30, 1986, D1; Michael Isikoff, "Space Official Criticizes Probe," *Washington Post*, March 15, 1986, A1.

7. Joseph P. Allen, interview by the author, Washington, DC, April 2, 2013.

8. Wayne Hale, interview by the author, April 12, 2013; Storer Rowley, "NASA Wizards' Legendary Infallibility Blew Up with Shuttle," *Chicago Tribune*, March 2, 1986.

9. Haynes Johnson, "NASA's Decaying Public Image," *Washington Post*, February 12, 1986, A2.

10. Brian M. Duff to distribution, "Space Transportation System (STS) Public Affairs Contingency Plan," March 13, 1983, with attached subject plan, folder 18179, NHRC/HQ; William Safire, "Handling Bad News," *New York Times*, January 30, 1986, A21; Thomas O'Toole, "After 27 Years of Openness, NASA Falls Silent," *Washington Post*, January 31, 1986, B2; Alex S. Jones, "Journalists Say NASA's Reticence Forced Them to Gather Data Elsewhere," *New York Times*, February 9, 1986; Matt Moffett and Laurie McGinley, "NASA, Once a Master of Publicity, Fumbles in Handling Shuttle Crisis," *Wall Street Journal*, February 14, 1986.

11. Lisa Malone, interview by the author, April 17, 2013.

12. NASA, "Statement by RADM Truly," news release 86-53, April 25, 1986, NHRC online, https://historydms.hq.nasa.gov/sites/default/files/DMS/e000021173.pdf; William V. Shannon, "Did NASA Conceal Astronauts' Fate?" *Philadelphia Inquirer*, April 24, 1986; Kathy Sawyer, "NASA Statement on Shuttle Deaths Assailed," *Washington Post*, August 2, 1986, A4.

13. Michael Cabbage, interview by the author, Washington, DC, June 3, 2013.

14. William Boot, "NASA and the Spellbound Press," *Columbia Journalism Review*, July/August 1986, 23. See, for example, Philip M. Boffey, "Space Agency Image: A Sudden Shattering," *New York Times*, February 5, 1986, A1; Stuart Diamond, "NASA Wasted Billions, Federal Audits Disclose," *New York Times*, April 23, 1986, A1; and Diamond, "NASA Cut or Delayed Safety Spending," *New York Times*, April 24, 1986, A1.

15. Johnson quoted in Sam A. Marshall, "NASA after *Challenger*: The Public Affairs Perspective," *Public Relations Journal* 42, no. 8 (1986): 23.

16. Presidential Commission on the Space Shuttle Challenger Accident, *Report of the Presidential Commission on the Space Shuttle Challenger Accident* (Washington, DC: US Government Printing Office, 1986).

17. Presidential Commission on the Space Shuttle Challenger Accident, *Report of the Presidential Commission on the Space Shuttle Challenger Accident*, vol. 2, appendix F.

18. Jessie Harris, memorandum for the record, "Science and Technology Committee Hearing to Consider the Challenger Accident Investigation Report, October 7, 1986," with attachment, NHRC/HQ.

19. Steven V. Roberts, "Congress Is Divided Over What to Do about NASA," *New York Times*, February 23, 1986.

20. John M. Logsdon, "The Space Shuttle Program: A Policy Failure?" *Science* 232, no. 4754 (1986): 1099–105; Stephen Budiansky with Robert Kaylor, "What's Wrong with America's Space Program," *U.S. News & World Report*, December 28, 1987–January 4, 1988, 32.

21. David Rosenbaum, "Should US Continue to Send People into Space?" *New York Times*, January 30, 1986, A18; Daniel S. Greenberg, "NASA's Misplaced Reliance on the Shuttle Program," *Los Angeles Times*, October 31, 1988; Thomas Gold, "Don't Send People into Space Unnecessarily," *New York Times*, September 28, 1987, A24.

22. Jon A. McBride to Lloyd Bentsen, April 11, 1988, folder 20116, NHRC/HQ.

23. James Fisher, "America Loves NASA, Budget Writers Don't," *Orlando Sentinel*, November 24, 1987; Jon D. Miller, "The Impact of the Challenger Accident on Public Attitudes toward the Space Program: A Report to the National Science Foundation," January 25, 1987, unnamed folder, box 16, Alan Ladwig unprocessed papers, NHRC/HQ.

24. Marshall, "NASA after *Challenger*," 24; statement of Jack Anderson in NASA Space Program, *Hearings Before the Subcommittees of the Comm. on Appropriations, United States Senate*, 99th Cong. 107–14 (1986).

25. See, for example, American Security Bank news release, January 28, 1986, folder 8521, NHRC/HQ; Associated Press, "Bank Shuttle Fund Nets $350,000," *Washington Post*, February 14, 1986; NASA, "US Space Foundation Outlines Challenger 7 Fund," news release 86-18, February 27, 1986, NHRC online, https://historydms.hq.nasa.gov/sites/default/files/DMS/e000021173.pdf.

26. NASA/Kennedy Space Center, "May Was Ninth Straight Record-Breaking Month at Spaceport USA," news release 87-64, June 1, 1987, folder 14098, NHRC/HQ; John Noble Wilford, "America's Future in Space: After the Challenger," *New York Times*, March 16, 1986.

27. National Commission on Space, *Pioneering the Space Frontier: The Report of the National Commission on Space* (New York: Bantam Books, 1986), 169; Marcia Smith, interview by the author, Arlington, VA, July 7, 2014.

28. "NASA: No Flight Plan," *Scientific American*, February 1987, 58; James C. Fletcher, "Excerpts from Remarks Prepared for Delivery: The Forum Club of Houston, February 19, 1988, Houston, TX," 7, folder 8977, NHRC/HQ.

29. Sally K. Ride, *Leadership and America's Future in Space* (Washington, DC: US Gov-

ernment Printing Office, 1987); Office of the Press Secretary, "Fact Sheet: Presidential Directive on National Space Policy," February 11, 1988, in *Organizing for Exploration*, vol. 1 of *Exploring the Unknown: Selected Documents in the History of the US Civil Space Program*, ed. John M. Logsdon, NASA SP-4407 (Washington, DC: US Government Printing Office, 1995), 605; Thor Hogan, *Mars Wars: The Rise and Fall of the Space Exploration Initiative*, NASA SP-2007-4410 (Washington, DC: US Government Printing Office, 2007).

30. Fletcher, "Excerpts from Remarks Prepared for Delivery," 3.

31. L/Director, Public Affairs (Shirley Green), to L/Associate Administrator for External Relations, "LF Objectives," February 10, 1987, folder 18180, NHRC/HQ; L/Deputy Associate Administrator for Communications (Shirley Green) to distribution, "Contingency Planning," November 25, 1987, with attachments, folder 18180, NHRC/HQ; NASA/Office of Space Flight, "National Space Transportation System Contingency Action Plan," July 1988, folder 7990, NHRC/HQ; June Malone, interview by the author, March 5, 2013.

32. Jon A. McBride to Daniel P. Moynihan, April 11, 1988; Jon A. McBride to Jimmy Hayes, April 6, 1988; both in folder 20116, NHRC/HQ.

33. Brian Welch, "Musings of an Unabashed Shuttle Apologist," *NASA Activities*, November/December 1990, 20–21, 24, folder 7990, NHRC/HQ; Suzanne Sataline, "NASA, Hamilton Standard Face Questions on Space Toilet's Price," *Hartford Courant*, February 24, 1993; June Malone, interview by the author, March 5, 2013.

34. Robert B. Seamans Jr. to Lewis Alumni, with attachment, "Fact Sheet," n.d., folder 6614, NHRC/HQ; NASA, "NASA Programs Generate More Than 300,000 Jobs, Study Shows," news release 89-94, folder 5715, NHRC/HQ; NASA, "Space Shuttle Research Reports," folder 7990, NHRC/HQ; Jack Anderson and Dale Van Atta, "Selling the Space Program," *Washington Post*, July 16, 1988; L/Associate Administrator for Communications (William Sheehan) to A/Aaron Cohen, "Role of Office of Communications in Moon/Mars Initiative," August 9, 1989, folder 8978, NHRC/HQ.

35. Kathy Sawyer, "Columbia Crew Shares TV Limelight," *Washington Post*, July 6, 1992, A17.

36. NASA, *Toward a Shared Vision: 1992 Town Meetings*, NASA NP-205 (Washington, DC: US Government Printing Office, 1993), 6.

37. NASA, "Applying NASA Shuttle Engine Test Findings May Save Airlines Millions in Fuel Costs," news release 96-254, December 12, 1996, folder 6901, NHRC/HQ; NASA Kennedy Space Center, "Space Shuttle Insulation to Be Commercially Produced for Use on Race Cars," news release 118-96, October 4, 1996, folder 8136, NHRC/HQ; NASA, "Spinoffs from the Space Shuttle Program," fact sheet FS-2000-01-007-HQ, 2000, folder 19748, NHRC/HQ.

38. Sarah C. Hoyt to Alan Ladwig, February 28, 2000, with attachment, "Space Awareness Alliance → Space Awareness Initiative, Frequently Asked Questions," folder "Space Awareness Alliance," box 14, Alan Ladwig unprocessed papers, NHRC/HQ; Paul

Hoversten, "NASA Supporters, Phone HOME, and Write to Congress," *USA Today*, March 22, 1996, folder 6620, NHRC/HQ; Rockwell, "We're Right on Course for a Rendezvous with Destiny" (advertisement), *Space News*, June 19–25, 1995; United Space Alliance, "Every Mission Improves Your World" (advertisement), *International Space Industry Report*, February 4, 1998, folder 11904, NHRC/HQ.

39. *America's Spaceport: John F. Kennedy Space Center*, [1986–1989?], pages unnumbered, folder 4636, NHRC/HQ; Jerry Hlass to NASA Speakers and Space Program Supporters, February 10, 1989, folder 18175, NHRC/HQ; James C. Fletcher to Patrick J. Leahy, May 27, 1988, NHRC/HQ.

40. L/Associate Administrator for Communications to distribution (William Sheehan), "Theme Message for NASA Spokesmen," February 29, 1988, folder 8977.

41. NASA, *Wings in Orbit: Scientific and Engineering Legacies of the Space Shuttle*, ed. Wayne Hale, Helen Lane, Gail Chapline, and Kamlesh Lula, NASA SP-2010-3409 (Washington, DC: US Government Printing Office, 2010), 36.

42. Hugh Harris, interview by Roger Launius, February 20, 2002, transcript, Kennedy Space Center History Project, Kennedy Space Center, FL.

43. Beth Beck, interview by the author, June 18, 2013.

44. "Delaware North Brings Vision to Spaceport USA," *Spaceport News*, May 19, 1995, 5, folder 14098, NHRC/HQ.

45. Richard Allen, interview by the author, Houston, TX, June 11, 2013.

46. Craig Covault, "'Earth Watch' Could Boost Interest in Shuttle Missions," *Aviation Week & Space Technology*, December 24, 1990, 53–54.

47. James Hartsfield, interview by the author, June 12, 2013; Sawyer, "Columbia Crew Shares TV Limelight."

48. NASA, "NASA Launches New Education Initiative with Disney Parks and Buzz Lightyear," news release 08-134, May 29, 2008, accessed May 2, 2023, http://www.nasa.gov/home/hqnews/2008/may/HQ_08134_Buzz_Lightyear.html.

49. NASA, Office of the Administrator website, [1994–1995?], folder "nasa.gov," box 15, Alan Ladwig unprocessed papers, NHRC/HQ.

50. "World-Wide Response," *Countdown*, July/August 1993, 18–19, folder 6620, NHRC/HQ; Rob Pegoraro, "NASA: Off to a Flying Start," *Washington Post*, July 6, 1995; "STS-95 Launch to be Webcast; Citizens Get Live Answers via Internet," *Florida Today*, October 28, 1998; James Hartsfield, interview by the author, June 12, 2013.

51. NASA, Name the New Orbiter program entry packet, 1988, folder 19382, NHRC/HQ; NASA, "President Bush Names Replacement Orbiter 'Endeavour,'" news release 89-70, May 10, 1989, folder 19382, NHRC/HQ; NASA, "Space Shuttle Overview: Endeavour (OV-105)," accessed May 2, 2023, http://www.nasa.gov/centers/kennedy/shuttleoperations/orbiters/endeavour-info.html.

52. NASA, "Endeavour Fund to Be Used for Teacher Fellowships," news release 91-

64, April 25, 1991, folder 5849, NHRC/HQ; NASA, "NASA Administrator Invites Local Students to Shuttle Launches," news release 90-48, April 2, 1990, folder 7990, NHRC/HQ; NASA, "Space Classroom—Assignment: The Stars," August 28, 1990, folder 5849, NHRC/HQ.

53. Presidential Commission on the Space Shuttle Challenger Accident, *Report of the Presidential Commission on the Space Shuttle Challenger Accident*.

54. Advisory Committee on the Future of the US Space Program, *Report of the Advisory Committee on the Future of the U.S. Space Program* (Washington, DC: US Government Printing Office, 1990).

55. "Bipartisan Call for Leadership to Save Space Program," August 8, 1986, folder 1570, NHRC/HQ; NASA, *Report to the President: Actions to Implement the Recommendations of The Presidential Commission on the Space Shuttle Challenger Accident, Executive Summary* (Washington, DC: US Government Printing Office, 1986), 4; President Ronald Reagan, "Statement by the President," August 15, 1986, in *Accessing Space*, vol. 4 of Logsdon, *Exploring the Unknown*, 383; The White House, Fact Sheet, NSDD-254, "United States Space Launch Strategy," December 27, 1986, in Logsdon, *Accessing Space*, 384–85.

56. Robert Bazell, "Shuttlecrock," *New Republic*, December 26, 1988, 13–14. The air force had received authority to contract with Martin Marietta Corporation to design the *Titan IV* in 1985.

57. James M. Beggs, interview by the author, April 8, 2013.

58. James M. Beggs, interview by Kevin M. Rusnak, Bethesda, MD, March 7, 2002, transcript, NASA Oral History, NHRC, Washington, DC, https://historycollection.jsc.nasa.gov/JSCHistoryPortal/history/oral_histories/NASA_HQ/Administrators/Beggs-JM/BeggsJM_3-7-02.htm.

59. Beggs, interview by the author, April 8, 2013.

60. Richard H. Truly to Sister Mary Carroll McCaffrey, July 29, 1987, folder 19759, NHRC/HQ; Charlie Jean, "Artist's Little World to Fly on Discovery," *Orlando Sentinel*, February 20, 1989; Nonscientific Payloads, 53 Fed. Reg. 47949 (November 29, 1988) (amending 14 C.F.R. pt. 1214).

61. NASA, "Reach for the Stars through NASA's Space Science Student Involvement Program," fact sheet, July 30, 1991, folder 5849, NHRC/HQ; NASA/National Science Teachers Association, *Everyone's a Winner with SSIP!*, EP-917, June 1994, folder 19816, NHRC/HQ.

62. Charles D. Walker, interview by Sandra Johnson, Springfield, VA, November 7, 2006, transcript, Johnson Space Center Oral History Project, Houston TX, https://historycollection.jsc.nasa.gov/JSCHistoryPortal/history/oral_histories/WalkerCD/WalkerCD_11-7-06.pdf; W. Michael Hawes, interview by the author, Washington, DC, July 15, 2014.

63. Cass Peterson, "Refocusing the Shuttle toward Science," *Washington Post*, Sep-

tember 26, 1988, A4; Dale D. Myers to Officials-in-Charge of Headquarters Office, Directors, NASA Field Installations, Director, Jet Propulsion Laboratory, "Secondary STS Payloads," June 5, 1987, folder 8183, NHRC/HQ; James K. Asker, "NASA Hikes Price for 'GAS' Payloads," *Aviation Week & Space Technology*, October 26, 1992, 73–74; Liz Tucci, "NASA Price Hike for Small Shuttle Users Cuts Demand," *Space News*, June 7–13, 1993, 22.

64. Asker, "NASA Hikes Price for 'GAS' Payloads"; Liz Tucci, "Shuttle GAS Canisters Program Running on Empty," *Space News*, October 11–17, 1993, 9; NASA, "NASA Reopens Reservations Queue for Get Away Specials," news release 92-174, October 20, 1992, folder 8183, NHRC/HQ; Use of Small Self-Contained Payloads, 57 Fed. Reg. 41077 (September 9, 1992) (amending 14 C.F.R. pt. 1214); Use of Small Self-Contained Payloads, 56 Fed. Reg. 47146 (September 18, 1991) (amending 14 C.F.R. pt. 1214).

65. Special Policy on Small Self-Contained Payloads by Domestic Educational Institutions, 57 Fed. Reg. 61794 (December 29, 1992) (amending 14 C.F.R. pt. 1214); Kathy Sawyer, "NASA's 'Get Away Special': Prices Soar Along with the Payload," *Washington Post*, November 19, 1991, A19.

66. A fourth proposed satellite was cancelled following the 2003 STS-107/*Columbia* accident.

67. NASA, *Wings in Orbit*, 474.

68. Sally K. Ride, interview by Rebecca Wright, San Antonio, TX, December 6, 2012, transcript, Johnson Space Center Oral History Project, Houston, TX, https://historycollection.jsc.nasa.gov/JSCHistoryPortal/history/oral_histories/RideSK/RideSK_12-6-02.htm.

69. NASA, *Wings in Orbit*, 475.

70. George W. S. Abbey, interview by the author, Houston, TX, June 12, 2013.

71. Seth Borenstein, "Space for Rent: Shuttle Columbia Could Take Advertising to New Heights," *Orlando Sentinel*, May 15, 1997; United Space Alliance, "Every Mission Improves Your World."

72. Alan M. Ladwig, "Administrator's Policy Review: National Policies and Statutes, and the Execution of NASA's Space Transportation Implementation Plan with Particular Attention to the Safe and Efficient Use of the Space Shuttle," February 10, 1998, folder 17980, NHRC/HQ.

73. Commercial Space Act of 1998, Pub. L. No. 105-303 112 Stat. 2843 (1998).

74. NASA, "NASA, Dreamtime Partnership Propels Space Information Age to New Heights," news release 00-87, June 2, 2000, accessed May 2, 2023, http://www.nasa.gov/home/hqnews/2000/00-087.txt.

75. Charles M. Chafer to Daniel S. Goldin, January 21, 1998; NASA Policy Directive 8870, "NASA Policy for the Flight and Disposal in Space of Human or Animal Remains" (draft), n.d.; both in folder 5382, NHRC/HQ.

76. Amanda Orion, "Ads in Space," *Fox News*, October 14, 1999; Bob Garfield, "Should

NASA Sell Ads?" *Air & Space*, June/July 2000, 74–75; Dee Ann Divis, "NASA Reviewing Sponsorship Possibilities," *Aviation Now*, September 5, 2000; all in folder 11904, NHRC/HQ.

77. NASA, "NASA Announces 2-Year Astronaut Selection Cycle," news release 88-63, May 11, 1988, folder 8929, NHRC/HQ.

78. Hale, interview by the author, April 12, 2013.

79. Robert W. Brown and Barbara R. Morgan, "Teacher in Space Program: The Challenge to Education in the Space Age," paper presented at the 37th Congress of the International Astronautical Federation, Innsbruck, Austria, October 4–11, 1986, folder 8076, NHRC/HQ; NASA, "Teacher in Space Program to Continue," news release 86-12, February 12, 1986, NHRC online, accessed May 2, 2023, https://historydms.hq.nasa.gov/sites/default/files/DMS/e000021173.pdf.

80. Ronald Reagan, "Address to the Nation on the Explosion of the Space Shuttle Challenger," January 28, 1986, Ronald Reagan Presidential Library and Museum, https://www.reaganlibrary.gov/archives/speech/address-nation-explosion-space-shuttle-challenger.

81. Reagan quoted in Bernard Weinraub, "Reagan Sees More Teachers in Space," *New York Times*, February 8, 1986, 17.

82. William R. Graham, "Remarks Prepared for Delivery: Press Briefing, Washington, D.C.," February 13, 1986, folder 8072, NHRC/HQ; Walter Pincus, "Space Flights by Teachers to Continue," *Washington Post*, February 14, 1986, A14.

83. Stewart and Glenn quoted in Walter Pincus, "NASA's Push to Put Citizen in Space Overtook Fully 'Operational' Shuttle," *Washington Post*, March 5, 1986, A8; Ernest F. Hollings, "Disaster Showed Delay Is Needed," *San Bernardino County (CA) Sun*, June 29, 1986, 54.

84. "Civilians-in-Space on Back Burner," *Defense Daily*, May 16, 1986, 90.

85. LE/Director, Educational Affairs Division (Robert W. Brown) to L/Associate Administrator for External Relations, "Background Information for Civilians in Space Policy Deliberations," March 20, 1987; LF/Director, Public Affairs (Shirley Green) to L/Associate Administrator for External Relations, "Civilians in Space Policy Deliberations," March 23, 1987; both in folder 8074, NHRC/HQ.

86. Association of Schools of Journalism and Mass Communication press release, May 14, 1986; Ann Bradley, NASA Associate Deputy Administrator, to Robert Hoskins, President, Association of Schools of Journalism and Mass Communication, July 1, 1986; both in folder 19754, NHRC/HQ. James C. Fletcher to Don Fuqua, September 26, 1986, folder 19759, NHRC/HQ.

87. Payload Specialists for NASA or NASA-Related Payloads, 54 Fed. Reg. 48587 (November 24, 1989) (amending 14 C.F.R. pt. 1214).

88. T. R. Reid, "Japanese Journalist in Orbit," *Washington Post*, December 6, 1990, D1; Leonard David, "Interest Revived in Citizens in Space," *Space News*, December 10–16, 1990, 4, 21.

89. Lynn W. Glass and Wendell Mohling, National Science Teachers Association, to President George Bush, July 26, 1991, folder 19755, NHRC/HQ; NASA, "NASA Administrator Supports Teaching from Space," news release 92-40, March 26, 1992, folder 19755, NHRC/HQ.

90. House Comm. on Science, Space, and Technology, National Aeronautics and Space Administration Multiyear Authorization Act of 1992, H.R. Rep. 102-500, at 59 (1992); Daniel Goldin, "On Risks Inherent in Exploration," *Los Angeles Times*, May 21, 1993; Alan Ladwig, interview by the author, December 28, 2010.

91. President Bill Clinton to Larry LaRocco, March 22, 1993, folder 19755, NHRC/HQ; Dirk Kempthorne to Daniel Goldin, August 10, 1994, folder 8076, NHRC/HQ.

92. See, for example, Ed Wardwell to Dan Goldin, [1994?]; Jerry Stoicheff to Dan Goldin, [1994?]; and Monica E. Beaudoin to Daniel Goldin, July 8, 1994, with attached petition from members of the National Education Association; all in folder 20198, NHRC/HQ.

93. Barbara Morgan to Dan Goldin, April 17, 1994; Goldin to Morgan, June 6, 1994; both in folder 19756, NHRC/HQ.

94. "The Flight of a Teacher in Space," August 25, 1994 (draft), folder 19755, NHRC/HQ.

95. Seth Borenstein, "Glenn Interested in Another Trip to Space," *Orlando Sentinel*, June 21, 1997; Paul Barton, "Sen. John Glenn May Return to Space in the Name of Medical Research," *Florida Today*, July 21, 1997; Francis X. Donnelly, "John Glenn: A Hero for the Ages Hopes for One Last Launch," *Florida Today*, September 28, 1997.

96. James Pura, "Sen. Glenn Shuttle Flight Most Expensive Congressional Junket Ever," Space Frontier Foundation, January 15, 1998, accessed May 29, 2023, https://web.archive.org/web/20150929044625/https://spacefrontier.org/1998/01/sen-glenn-shuttle-flight-most-expensive-congressional-junket-ever/; National Space Society, "Statement on Possibility of John Glenn Shuttle Flight," January 15, 1998, accessed December 10, 2016, https://web.archive.org/web/20161210142020/http://www.nss.org/news/releases/release24.html.

97. Daniel S. Goldin, "Announcement of NASA's Decision to Fly John Glenn," January 16, 1998, NHRC online, accessed May 2, 2023, https://historydms.hq.nasa.gov/sites/default/files/DMS/e000013684.pdf; NASA, "Media Availability with First Educator/Mission Specialist in Houston, Tuesday," news release N98-6, January 16, 1998, accessed May 2, 2023, http://www.nasa.gov/home/hqnews/note2edt/1998/n98-006.txt.

98. Sean O'Keefe, "Pioneering the Future," address at the Maxwell School of Citizenship and Public Affairs, Syracuse University, April 12, 2002, NASA Historical Reference Center online, https://historydms.hq.nasa.gov/sites/default/files/DMS/e000040280.pdf.

99. NASA, "Educator Mission Specialist Program Plan," September 13, 2002 [handwritten], folder 19749, NHRC/HQ.

100. NASA Educator Astronaut Program Office, Code N-EAP, "Positive Comments Received: Post Columbia Tragedy," February 7, 2003, folder 19748, NHRC/HQ; NASA,

"NASA Introduces the Next Generation of Explorers," news release 04-152, May 6, 2004, http://www.nasa.gov/home/hqnews/2004/may/HQ_04152_space_day_ascan.html.

101. Borenstein, "Space for Rent."

102. Hawes, interview interview by the author, July 15, 2014.

103. Ralph M. Hall to Dan Goldin, February 6, 2001, "Rep. Hall Asks NASA about Station Tourist," Feburary 8, 2001, SpaceRef.com, https://spaceref.com/press-release/rep-hall-asks-nasa-about-station-tourist/; Richard Stenger, "John Glenn: Space Tourist Cheapening Alpha," CNN.com, May 3, 2001, accessed May 2, 2023, http://edition.cnn.com/2001/TECH/space/05/03/space.day/.

104. Paul Recer, "Space Tourist Dennis Tito Used 'Wrong Way' to Get to Orbit, Says NASA Chief," *Ludington (MI) Daily News*, May 3, 2001, B12; Hawes, interview by the author, July 15, 2014; Alan Ladwig, interview by the author, January 31, 2013.

105. Tito quoted in Alan Boyle, "Tourist Wants to Work with NASA," MSNBC.com, May 18, 2001, folder 17247, NHRC/HQ.

106. Dennis Tito, "Expanding the Dream of Human Space Flight," written testimony prepared for the US House of Representatives, Committee of Science, Subcommittee on Space and Aeronautics, hearing on space tourism, June 26, 2001, SpaceRef.com, https://spaceref.com/press-release/testimony-by-dennis-tito-before-the-house-subcommittee-on-space-and-aeronautics/.

107. W. Michael Hawes, written testimony prepared for the US House of Representatives, Committee of Science, Subcommittee on Space and Aeronautics, hearing on space tourism, June 26, 2001, "Testimony by W. Michael Hawes on 'Space Tourism' before the House Subcommittee on Space and Aeronautics," SpaceRef.com, June 26, 2001, https://spaceref.com/press-release/testimony-by-w-michael-hawes-on-space-tourism-before-the-house-subcommittee-on-space-and-aeronautics/.

Chapter 8: Commemorating the People's Spaceship

1. National Academies of Sciences, Engineering, and Medicine, *Upgrading the Space Shuttle* (Washington, DC: National Academies Press, 1999), 1.

2. Aerospace Safety Advisory Panel, *Annual Report for 1998* (Washington, DC: US Government Printing Office, February 1999); Aerospace Safety Advisory Panel, *Annual Report for 1999* (Washington, DC: US Government Printing Office, 2000), 15.

3. The mission's inclusion of the first Israeli payload specialist, Ilan Ramon, nonetheless attracted some additional media attention.

4. Michael Cabbage, interview by the author, June 3, 2013.

5. Lisa Malone, interview by the author, April 17, 2013.

6. Jonathan Krezel, interview by the author, Washington, DC, March 15, 2013; NASA, *NASA's Implementation Plan for Space Shuttle Return to Flight and Beyond*, May 15, 2007, E1-E3, https://www.nasa.gov/wp-content/uploads/2015/01/178101main_rtfip_final_200705.pdf.

7. Jeffrey M. Jones, "Support for Space Program Funding High by Historical Standards," August 19, 2003, Gallup News Service, http://www.gallup.com/poll/9082/Support-Space-Program-Funding-High-Historical-Standards.aspx.

8. Milt Heflin, interview by the author, May 3, 2013.

9. See, for example, James Glanz with William J. Broad, "Has the Space Shuttle Turned into NASA's '76 Dodge Dart?" *New York Times*, January 27, 2004.

10. "The Call of Distant Worlds," *New York Times*, February 9, 2003.

11. Columbia Accident Investigation Board, *Report Volume I* (Washington, DC: US Government Printing Office, 2003), 209, 23.

12. President George W. Bush, *A Renewed Spirit of Discovery: The President's Vision for US Space Exploration*, January 2004, http://history.nasa.gov/renewedspiritofdiscovery.pdf.

13. NE/Manager, Flight Projects Office (Debbie Brown Biggs), and NT/Program Executive, Technology and Products Office (Shelley Canright), to N/Associate Administrator for Education, August 6, 2003; NASA, "Join the Earth Crew" webpage, 1993. Both documents located in folder 19748, NASA Historical Reference Collection, NASA Headquarters, Washington, DC (hereafter cited as NHRC/HQ).

14. Beth Beck, interview by the author, May 16, 2013; NASA, "NASA Hosts Launch and Mission Tweetups for Next Space Shuttle Mission," news release 10-088, April 14, 2010, http://www.nasa.gov/home/hqnews/2010/apr/HQ_10-088_STS-132_Tweetups.html.

15. NASA, "NASA Announces STS-133 Wakeup Song Winners; Face In Space Totals," news release 11-055, February 25, 2011, http://www.nasa.gov/home/hqnews/2011/feb/HQ_11-055_Top_40.html; NASA, "NASA Opens Voting for Original Songs to Awaken Next Shuttle Crew," media advisory 11-068, March 29, 2011, http://www.nasa.gov/home/hqnews/2011/mar/HQ_M11-068_Original_Song_Contest.html.

16. James Hull, interview by the author, Washington, DC, January 31, 2013; NASA, "NASA Is Expanding Offer for Space Shuttle Tiles and Food," news release 12-304, September 4, 2012, http://www.nasa.gov/home/hqnews/2012/sep/HQ_12-304_NASA_Food_Tiles_schools.html.

17. NASA, Office of Inspector General, *Review of NASA's Selection of Display Locations for the Space Shuttle Orbiters*, August 25, 2011, 3–16, http://oig.nasa.gov/audits/reports/FY11/Review_NASAs_Selection_Display_Locations.pdf.

18. CBS News, "The Space Shuttle Program," July 8, 2011, http://www.cbsnews.com/htdocs/pdf/JUN11B_1.PDF; CNN/ORC poll, July 21, 2011, http://i2.cdn.turner.com/cnn/2011/images/07/21/poll.july21.pdf; Pew Research Center, "Majority Sees U.S. Leadership in Space as Essential," July 5, 2011, http://www.people-press.org/2011/07/05/majority-sees-u-s-leadership-in-space-as-essential/.

19. Roger D. Launius, "Public Opinion Polls and Perceptions of US Human Space Flight," *Space Policy* 19, no. 3 (2003): 167–68.

20. CBS News, "Space Shuttle Program"; David Gergen and Michael Zuckerman,

"What Space Shuttle Discovery Has Inspired in Us," CNN.com, April 20, 2012, http://www.cnn.com/2012/04/20/opinion/zuckerman-gergen-discovery/; Adam Naids, interview by the author, January 23, 2013.

Conclusion

1. NASA, *NASA's Moon to Mars Strategy and Objectives Development*, 2023, https://www.nasa.gov/wp-content/uploads/2023/04/m2m_strategy_and_objectives_development.pdf.

2. Morning Consult Pro, "Nearly Half the Public Wants the U.S. to Maintain Its Space Dominance. Appetite for Space Exploration Is a Different Story," February 25, 2021, https://pro.morningconsult.com/articles/space-force-travel-exploration-poll.

3. Leonard David, "Equal Access to Space: New Study Investigates How to Get More 'Parastronauts' Aloft," Space.com, December 31, 2021, https://www.space.com/inclusive-human-spaceflight-parastronaut-study.

4. Zachary Pirtle, Katherine McBrayer, and Alyse Beauchemin, *Artemis, Society and Ethics: Synthesis from a Workshop*, Report ID 20230012799, NASA, Office of Technology, Policy, and Strategy, September 21, 2023, https://ntrs.nasa.gov/api/citations/20230012799/downloads/otps_artemis_ethics_and_society_report_final_v3.pdf. The author attended this workshop.

5. NASA, *Informing NASA's Asteroid Initiative: A Citizen Forum*, August 28, 2014, http://www.nasa.gov/content/informing-nasa-s-asteroid-initiative-a-citizen-forum/. The author was a part of the planning process for these forums.

6. Amy Paige Kaminski, "Can the Demos Make a Difference? Prospects for Participatory Democracy in Shaping the Future Course of US Space Exploration," *Space Policy* 28, no. 4 (2012): 225–33.

Bibliography

Archives

Johnson Space Center History Collection, University of Houston–Clear Lake, Houston, TX

Kennedy Space Center History Project. Kennedy Space Center, FL. Online at https://ks-coralhistory.ksc.nasa.gov/. Site discontinued; accessible through Internet Archive

NASA Historical Reference Collection, NASA Headquarters, Washington, DC

National Air and Space Museum Archives, Smithsonian Institution, Washington, DC. Online at https://airandspace.si.edu/archives

Reagan Library Online Archive. Ronald Reagan Presidential Library and Museum, Simi Valley, CA. https://www.reaganlibrary.gov

Primary and Secondary Sources

Abelson, Philip H. "Manned Lunar Landing." *Science* 140, no. 3564 (1963): 267.

Advisory Committee on the Future of the US Space Program. *Report of the Advisory Committee on the Future of the U.S. Space Program.* Washington, DC: US Government Printing Office, 1990.

Aerospace Safety Advisory Panel. *Annual Report for 1998.* Washington, DC: US Government Printing Office, 1999.

Aerospace Safety Advisory Panel. *Annual Report for 1999.* Washington, DC: US Government Printing Office, 2000.

Atkinson, Joseph D., Jr., and Jay M. Shafritz. *The Real Stuff: A History of NASA's Astronaut Recruitment Program.* New York: Praeger, 1985.

Beck, Ulrich. *Risk Society: Towards a New Modernity.* Translated by Mark Ritter. London: Sage, 1992.

Billings, Linda. "Fifty Years of NASA and the Public: What NASA? What Publics?" In *NASA's First 50 Years: Historical Perspectives,* edited by Steven J. Dick, 151–81. NASA SP-2010-4704. Washington, DC: US Government Printing Office, 2010.

Bromberg, Joan Lisa. *NASA and the Space Industry.* Baltimore, MD: Johns Hopkins University Press, 1999.

Burrows, William E. *This New Ocean: The Story of the First Space Age.* New York: Random House, 1998.

Bush, Vannevar. *Science, The Endless Frontier: A Report to the President.* Washington, DC: Government Printing Office, 1945.

Byrnes, Mark E. *Politics and Space: Image Making by NASA*. Westport, CT: Praeger, 1994.

Callon, Michel. "Some Elements of a Sociology of Translation: Domestication of the Scallops and the Fishermen of St. Brieuc Bay." In *Power, Action, and Belief: A New Sociology of Knowledge?*, edited by John Law, 196–233. London: Routledge and Kegan Paul, 1986.

Collins, H. M., and Robert Evans. "The Third Wave of Science Studies: Studies of Expertise and Experience." *Social Studies of Science* 32, no. 2 (2002): 235–96.

Columbia Accident Investigation Board. *Report Volume I*. Washington, DC: US Government Printing Office, 2003.

Compton, W. David, and Charles D. Benson. *Living and Working in Space: A History of Skylab*. NASA SP-4208. Washington, DC: US Government Printing Office, 1983.

Compton, William D. *Where No Man Has Gone Before: A History of Apollo Lunar Exploration Missions*. NASA SP-4214. Washington, DC: Government Printing Office, 1989.

Cooper, Henry S. F. *Before Lift-Off: The Making of a Space Shuttle Crew*. Baltimore, MD: Johns Hopkins University Press, 1987.

Corn, Joseph J. *The Winged Gospel: America's Romance with Aviation, 1900–1950*. New York: Oxford University Press, 1983.

Croft, Melvin, and John Youskauskas. *Come Fly with Us: NASA's Payload Specialist Program*. Lincoln: University of Nebraska Press, 2019.

Cunningham, Walter. *The All-American Boys*. New York: Macmillan, 1977.

Dethloff, Henry C. *Suddenly, Tomorrow Came . . . : A History of the Johnson Space Center*. NASA SP-4307. Washington, DC: US Government Printing Office, 1993.

Dewey, John. *The Public and Its Problems*. New York: Henry Holt, 1927.

Dubbs, Chris, and Emeline Paat-Dahlstrom. *Realizing Tomorrow: The Path to Private Spaceflight*. Lincoln: University of Nebraska Press, 2011.

Ehrlichman, John. *Witness to Power: The Nixon Years*. New York: Simon and Schuster, 1982.

Eiseley, Loren. *The Invisible Pyramid*. New York: Charles Scribner's Sons, 1970.

Etzioni, Amitai. *The Moon-doggle: Domestic and International Implications of the Space Race*. Garden City, NY: Doubleday, 1964.

Evans, Michael K. *The Economic Impact of NASA R&D Spending*. Final Report. Executive Summary. Bala Cynwyd, PA: Chase Economic Associates, 1976.

Ezrahi, Yaron. *The Descent of Icarus: Science and the Transformation of Contemporary Democracy*. Cambridge, MA: Harvard University Press, 1990.

Feenberg, Andrew. *Questioning Technology*. New York: Routledge, 1999.

Funtowicz, Silvio O., and Jerome R. Ravetz. "Science for the Post-normal Age." *Futures* 25, no. 7 (1993): 739–55.

Glennan, T. Keith. *The Birth of NASA: The Diary of T. Keith Glennan*. Edited by J. D. Hunley. NASA SP-4105. Washington, DC: US Government Printing Office, 1993.

Grey, Jerry. *Enterprise*. New York: William Morrow, 1979.

Guston, David H. "Forget Politicizing Science: Let's Democratize Science!" *Issues in Science and Technology* 21, no. 1 (2004): 25–28.

Harris, Gordon L. *Selling Uncle Sam*. Hicksville, NY: Exposition, 1976.

Heiss, Klaus P., and Oskar Morgenstern. *Mathematica Economic Analysis of the Space Shuttle System*. Princeton, NJ: Mathematica, 1972.

Heppenheimer, T. A. *History of the Space Shuttle*. Vol. 2, *Development of the Shuttle, 1972–1981*. Washington, DC: Smithsonian Institution Press, 2002.

Heppenheimer, T. A. *The Space Shuttle Decision: NASA's Search for a Reusable Space Vehicle*. NASA SP-4221. Washington, DC: US Government Printing Office, 1999.

Hersch, Matthew H. *Inventing the American Astronaut*. New York: Palgrave Macmillan, 2012.

Hoban, Francis T. *Where Do You Go after You've Been to the Moon? A Case Study of NASA's Pioneer Effort at Change*. Malabar, FL: Krieger, 1997.

Hogan, Thor. *Mars Wars: The Rise and Fall of the Space Exploration Initiative*. NASA SP-2007-4410. Washington, DC: US Government Printing Office, 2007.

Hubbard, Barbara Marx. *The Hunger of Eve*. Harrisburg, PA: Stackpole, 1976.

Igo, Sarah E. *The Averaged American: Surveys, Citizens, and the Making of a Mass Public*. Cambridge, MA: Harvard University Press, 2007.

Jasanoff, Sheila. "Technologies of Humility: Citizen Participation in Governing Science." *Minerva* 41, no. 3 (2003): 223–44.

Jasanoff, Sheila, and Sang-Hyun Kim. "Containing the Atom: Sociotechnical Imaginaries and Nuclear Power in the United States and South Korea." *Minerva* 47, no. 2 (2009): 119–46.

Kaminski, Amy Paige. "Can the Demos Make a Difference? Prospects for Participatory Democracy in Shaping the Future Course of US Space Exploration." *Space Policy* 28, no. 4 (2012): 225–33.

Kaminski, Amy Paige. "Explorers We? The Making, Unmaking, and Public Involvement Legacy of NASA's Space Flight Participant Program." *Quest: The History of Space Flight Quarterly* 20, no. 1 (2013): 12–21.

Kauffman, James L. *Selling Outer Space: Kennedy, the Media, and Funding for Project Apollo, 1961–1963*. Tuscaloosa: University of Alabama Press, 1994.

Kay, W. D. *Can Democracies Fly in Space? The Challenge of Revitalizing the U.S. Space Program*. Westport, CT: Praeger, 1995.

Kelly, Ellen I. "The Young Astronauts." *Aerospace: Official Publication of the Aerospace Industries Association*, Summer 1985, 14–16.

Kevles, Bettyann Holtzmann. *Almost Heaven: The Story of Women in Space*. Cambridge: MIT Press, 2006.

Kitcher, Philip. *Science, Truth, and Democracy*. New York: Oxford University Press, 2001.

Krugman, Herbert E. "Public Attitudes toward the Apollo Space Program, 1965-1975." *Journal of Communication* 27, no. 4 (1977): 87-93.

Ladwig, Alan. *See You in Orbit? Our Dream of Spaceflight.* Falls Church, VA: To Orbit Productions, 2019.

Lahsen, Myanna. "Technocracy, Democracy, and U.S. Climate Politics: The Need for Demarcations." *Science, Technology, and Human Values* 30, no. 1 (2005): 137-69.

Latour, Bruno. "Give Me a Laboratory and I Will Raise the World." In *Science Observed: Perspectives on the Social Study of Science*, edited by Karin D. Knorr-Cetina and Michael Mulkay, 141-69. Beverly Hills, CA: Sage, 1983.

Latour, Bruno. *Science in Action: How to Follow Scientists and Engineers through Society.* Cambridge, MA: Harvard University Press, 1987.

Latour, Bruno. *We Have Never Been Modern.* Translated by Catherine Porter. Cambridge, MA: Harvard University Press, 1993.

Launius, Roger D. "First Steps into Space: Projects Mercury and Gemini." In *Exploring the Unknown: Selected Documents in the History of the US Civil Space Program*, vol. VII: *Human Space Flight: Projects Mercury, Gemini, and Apollo*, edited by John M. Logsdon, 1-48. NASA SP-2008-4407. Washington, DC: US Government Printing Office, 2008.

Launius, Roger D. "Heroes in a Vacuum: The Apollo Astronaut as Cultural Icon." *Florida Historical Quarterly* 87, no. 2 (2008): 174-209.

Launius, Roger D. "Prelude to the Space Age." In *Exploring the Unknown: Selected Documents in the History of the US Civil Space Program*, vol. I: *Organizing for Exploration*, edited by John M. Logsdon, 1-21. NASA SP-4407. Washington, DC: US Government Printing Office, 1995.

Launius, Roger D. "Public Opinion Polls and Perceptions of US Human Space Flight." *Space Policy* 19, no. 3 (2003): 163-75.

Launius, Roger D. "The Strange Career of the American Spaceplane: The Long History of Wings and Wheels in Human Space Operations." *Centaurus: An International Journal of the History of Science and Its Cultural Aspects* 55, no. 4 (2013): 412-32.

Launius, Roger D., and Howard E. McCurdy, eds. *Spaceflight and the Myth of Presidential Leadership.* Urbana: University of Illinois Press, 1997.

Law, John. "Technology and Heterogeneous Engineering: The Case of Portuguese Expansion." In *The Social Construction of Technological Systems: New Directions in the Sociology and History of Technology*, edited by Wiebe E. Bijker, Thomas P. Hughes, and Trevor Pinch, 111-34. Cambridge: MIT Press, 1989.

Lengwiler, Martin. "Participatory Approaches in Science and Technology: Historical Origins and Current Practices in Critical Perspective." *Science, Technology, and Human Values* 33, no. 2 (2008): 186-200.

Lewenstein, Bruce V. "NASA and the Public Understanding of Space Science." *Journal of the British Interplanetary Society* 46 (1993): 251-54.

Lewenstein, Bruce V. "'Public Understanding of Science' in America, 1945–1965." PhD diss., University of Pennsylvania, 1987.

Lippmann, Walter. *The Phantom Public*. New York: Harcourt, Brace, 1925.

Logsdon, John M., ed. *Accessing Space*. Vol. 4 of *Exploring the Unknown: Selected Documents in the History of the US Civil Space Program*. NASA SP-4407. Washington, DC: US Government Printing Office, 1999.

Logsdon, John M. *After Apollo? Richard Nixon and the American Space Program*. New York: Palgrave Macmillan, 2015.

Logsdon, John M. *The Decision to Go to the Moon: Project Apollo and the National Interest*. Cambridge: MIT Press, 1970.

Logsdon, John M. "The Evolution of US Space Policy and Plans." In *Exploring the Unknown: Selected Documents in the History of the US Civil Space Program*, vol. I: *Organizing for Exploration*, edited by John M. Logsdon, 377–93. NASA SP-4407. Washington, DC: US Government Printing Office, 1995.

Logsdon, John M., ed. *Human Space Flight: Projects Mercury, Gemini, and Apollo*. Vol. 7 of *Exploring the Unknown: Selected Documents in the History of the US Civil Space Program*. NASA SP-2008-4407. Washington, DC: US Government Printing Office, 2008.

Logsdon, John M. *John F. Kennedy and the Race to the Moon*. New York: Palgrave Macmillan, 2010.

Logsdon, John M., ed. *Organizing for Exploration*. Vol. 1 of *Exploring the Unknown: Selected Documents in the History of the US Civil Space Program*. NASA SP-4407. Washington, DC: US Government Printing Office, 1995.

Logsdon, John M. "The Space Shuttle Program: A Policy Failure?" *Science* 232, no. 4754 (1986): 1099–105.

Mark, Hans. *The Space Station, A Personal Journey*. Durham, NC: Duke University Press, 1987.

Marshall, Sam A. "NASA after *Challenger*: The Public Affairs Perspective." *Public Relations Journal* 42, no. 8 (1986): 17–19, 22–24, 39.

Martin, Brian. "Strategies for Alternative Science." In *The New Political Sociology of Science: Institutions, Networks, and Power*, edited by Scott Frickel and Kelly Moore, 272–98. Madison: University of Wisconsin Press, 2006.

McCurdy, Howard E. *Inside NASA: High Technology and Organizational Change in the U.S. Space Program*. Baltimore, MD: Johns Hopkins University Press, 1993.

McCurdy, Howard E. *Space and the American Imagination*. Washington, DC: Smithsonian Institution Press, 1997.

McDougall, Walter A. . . . *The Heavens and the Earth: A Political History of the Space Age*. Baltimore, MD: Johns Hopkins University Press, 1985.

McQuaid, Kim. "Race, Gender, and Space Exploration: A Chapter in the Social History of the Space Age." *Journal of American Studies*, 41, no. 2 (2007): 405–44.

McQuaid, Kim. "Selling the Space Age: NASA and Earth's Environment, 1958–1990." *Environment and History* 12, no. 2 (2006): 127–63.

McQuaid, Kim. "Sputnik Reconsidered: Image and Reality in the Early Space Age." *Canadian Review of American Studies* 37, no. 3 (2007): 371–401.

Michael, Donald N. "The Beginning of the Space Age and American Public Opinion." *Public Opinion Quarterly* 24, no. 4 (1960): 573–82.

Michael, Donald N. "Society and Space Exploration." *Astronautics*, February 1958, 20–22, 88–90.

Michaud, Michael A. G. *Reaching for the High Frontier: The American Pro-space Movement, 1972–84.* New York: Praeger, 1986.

Midwest Research Institute. *Economic Impact of Stimulated Technological Activity.* MRI Project No. 3430-D. Final Report Summary. April 7, 1970–October 15, 1971.

NASA. *Astronautics and Aeronautics, 1969: Chronology on Science, Technology, and Policy.* NASA SP-4014. Washington, DC: US Government Printing Office, 1970.

NASA. *Astronautics and Aeronautics, 1978: A Chronology.* NASA SP-4023. Washington, DC: US Government Printing Office, 1986.

NASA. Division of Public Affairs. *Aboard the Space Shuttle.* By Florence S. Steinberg. NASA EP-169. Washington, DC: NASA, 1980.

NASA. *Exploring Space . . . Project Mercury.* Washington, DC: Government Printing Office, 1961.

NASA. Goddard Space Flight Center, Special Payloads Division. *Get Away Special . . . The First Ten Years.* Greenbelt, MD: NASA, 1989.

NASA. *Informing NASA's Asteroid Initiative: A Citizen Forum.* Washington, DC: NASA, 2014.

NASA. Langley Research Center. *69 Months in Space: A History of the First LDEF (Long Duration Exposure Facility).* NASA NP-149. Hampton, VA: NASA, 1992.

NASA. Lewis Research Center. *Launching a Dream: A Teachers Guide to a Simulated Space Shuttle Mission.* Washington, DC: US Government Printing Office, 1988.

NASA. *Man in Space: Space in the Seventies.* By Walter Froehlich. NASA EP-81. Washington, DC: US Government Printing Office, 1971.

NASA. *A Meeting with the Universe.* Edited by Bevan M. French and Stephen P. Moran. NASA EP-177. Washington, DC: NASA, 1981.

NASA. *NASA's Implementation Plan for Space Shuttle Return to Flight and Beyond.* Washington, DC: NASA, May 15, 2007.

NASA. *NASA's Moon to Mars Strategy and Objectives Development.* Washington, DC: NASA, 2023.

NASA. Office of Inspector General. *Review of NASA's Selection of Display Locations for the Space Shuttle Orbiters.* August 25, 2011.

NASA. Office of Space Science and Applications. Microgravity Science and Applications Division. *Microgravity . . . A New Tool for Basic and Applied Research in Space.* NASA EP-212. Washington, DC: US Government Printing Office, n.d.

NASA. Outlook for Space Study Group. *Outlook for Space: Report to the NASA Administrator.* Washington, DC: NASA, 1976.

NASA. *Report to the President: Actions to Implement the Recommendations of The Presidential Commission on the Space Shuttle Challenger Accident, Executive Summary.* Washington, DC: US Government Printing Office, 1986.

NASA. *SEEDS: A Celebration of Science.* NASA EP-281. Washington, DC: US Government Printing Office, 1991.

NASA. *Space Resources and Space Settlement.* NASA SP-428. Washington, DC: US Government Printing Office, 1979.

NASA. *Toward a Shared Vision: 1992 Town Meetings.* NASA NP-205. Washington, DC: US Government Printing Office, 1993.

NASA. *Wings in Orbit: Scientific and Engineering Legacies of the Space Shuttle.* Edited by Wayne Hale, Helen Lane, Gail Chapline, and Kamlesh Lula. NASA SP-2010-3409. Washington, DC: US Government Printing Office, 2010.

NASA. Marshall Space Flight Center. *Monodisperse Latex Reactor (MLR): A Materials Processing Space Shuttle Mid-Deck Payload.* By Dale M. Kornfeld. TM-86487. Washington, DC: US Government Printing Office, 1985.

NASA. Marshall Space Flight Center. *Skylab: Classroom in Space.* Edited by Lee B. Summerlin. NASA SP-401. Washington, DC: US Government Printing Office, 1977.

NASA. Marshall Space Flight Center. *Skylab Student Project Summary Description.* MS-FC-SL-73-3. Marshall Space Flight Center, AL: NASA, 1973.

National Academies of Sciences, Engineering, and Medicine. *Upgrading the Space Shuttle.* Washington, DC: National Academies Press, 1999.

National Commission on Excellence in Education. *A Nation at Risk: The Imperative for Educational Reform.* Washington, DC: US Government Printing Office, 1983.

National Commission on Space. *Pioneering the Space Frontier: The Report of the National Commission on Space.* New York: Bantam Books, 1986.

National Research Council. Space Science Board. *A Review of Space Research.* Washington, DC: National Academy Press, 1962.

National Research Council. Space Science Board. *Space Research: Directions for the Future.* Washington, DC: National Academy Press, 1966.

Neal, Valerie. *Spaceflight in the Shuttle Era and Beyond: Redefining Humanity's Purpose in Space.* New Haven, CT: Yale University Press with Smithsonian National Air and Space Museum, 2017.

Nelkin, Dorothy. "The Political Impact of Technical Expertise." *Social Studies of Science* 5, no. 1 (1975): 35–54.

Newell, Homer E. *Beyond the Atmosphere: Early Years of Space Science*. Washington, DC: US Government Printing Office, 1980.

Normyle, William J. "Manned Mission to Mars Opposed." *Aviation Week & Space Technology*, August 18, 1969, 16–17.

Nye, David E. *American Technological Sublime*. Cambridge: MIT Press, 1994.

Overholt, William, Anthony J. Wiener, and Doris Yokelson. *Implications of Public Opinion for Space Program Planning, 1980–2000*. Draft final report. HI-2219/2-RR. Croton-on-Hudson, NY: Hudson Institute, 1975.

Pace, Scott. "Engineering Design and Political Choice: The Space Shuttle, 1969–1972." Master's thesis, Massachusetts Institute of Technology, 1982.

Philpott, Jeffrey S. "From 'Major Malfunction' to 'Pulling Us into the Future': Rhetorical Transformation and the Reconstruction of Public Coherence in Immediate Public Responses to the Explosion of Space Shuttle *Challenger*." PhD diss., University of Washington, 1995.

Pielke, Roger A., Jr. "A Reappraisal of the Space Shuttle Programme." *Space Policy* 9, no. 2 (1993): 133–57.

Pirtle, Zachary, Katherine McBrayer, and Alyse Beauchemin. *Artemis, Society and Ethics: Synthesis from a Workshop*. Report ID 20230012799. NASA, Office of Technology, Policy, and Strategy, September 21, 2023, https://ntrs.nasa.gov/api/citations/20230012799/downloads/otps_artemis_ethics_and_society_report_final_v3.pdf

President's Commission on Implementation of United States Space Exploration Policy. *Report of the President's Commission on Implementation of United States Space Exploration Policy: A Journey to Inspire, Innovate, and Discover*. Washington, DC: US Government Printing Office, 2004.

Presidential Commission on the Space Shuttle Challenger Accident. *Report of the Presidential Commission on the Space Shuttle Challenger Accident*. Washington, DC: US Government Printing Office, 1986.

Price, Don K. *The Scientific Estate*. Cambridge, MA: Harvard University Press, 1967.

Ride, Sally K. *Leadership and America's Future in Space*. Washington, DC: US Government Printing Office, 1987.

Sclove, Richard. *Democracy and Technology*. New York: Guilford, 1995.

Scott, David Meerman, and Richard Jurek. *Marketing the Moon: The Selling of the Apollo Lunar Program*. Cambridge: MIT Press, 2014.

Scott, James C. *Seeing Like a State: How Certain Schemes to Improve the Human Condition Have Failed*. New Haven, CT: Yale University Press, 1999.

Shayler, David J., and Colin Burgess. *NASA's Scientist-Astronauts*. New York: Springer-Praxis, 2007.

Smith, Robert W. *The Space Telescope: A Study of NASA, Science, Technology, and Politics*. Cambridge: Cambridge University Press, 1989.

Space Task Group. *The Post-Apollo Space Program: Directions for the Future*, September 1969. In *Exploring the Unknown: Selected Documents in the History of the US Civil Space Program*, vol. 1: *Organizing for Exploration*, edited by John M. Logsdon, 522–43. NASA SP-4407. Washington, DC: US Government Printing Office, 1995.

Starr, Kristen A. "NASA's Hidden Power: NACA/NASA Public Relations and the Cold War, 1945–1967." PhD diss., Auburn University, 2008.

Taylor, L. B., Jr. *For All Mankind: America's Space Programs of the 1970s and Beyond*. New York: E. P. Dutton, 1974.

Vaughan, Diane. *The Challenger Launch Decision: Risky Technology, Culture, and Deviance at NASA*. Chicago: University of Chicago Press, 1996.

Weitekamp, Margaret A. *Right Stuff, Wrong Sex: America's First Women in Space Program*. Baltimore, MD: Johns Hopkins University Press, 2004.

Wiener, Anthony J., and B. Bruce-Briggs. *Contextual Planning for NASA: A Second Workbook of Alternative Future Environments for Mission Analysis*, Interim Report II. HI-1272/3-RR. Croton-on-Hudson, NY: Hudson Institute, 1971.

Wilcox, David A. "The Get Away Special Program: Year 2000 and Beyond." In *1999 Shuttle Small Payloads Symposium*, by NASA, edited by Gerard Daelemans and Frances L. Mosier, 91–95. NASA/CP-1999-209476. Greenbelt, MD: NASA, 1999.

Wilford, John Noble. "A Spacefaring People: Keynote Address." In *A Spacefaring People: Perspectives on Early Spaceflight*, edited by Alex Roland, 68–73. NASA SP-4405. Washington, DC: US Government Printing Office, 1985.

Winner, Langdon. "Do Artifacts Have Politics?" *Daedalus* 109, no. 1 (1980): 121–36.

Wolfe, Tom. *The Right Stuff*. New York: Picador, 1979.

Woods, Brian. "Artifacts, Revolutionaries, and Bureaucrats: The Sociotechnical Shaping of NASA's Space Shuttle." PhD diss., University of Edinburgh, 1998.

Wynne, Brian. "May the Sheep Safely Graze? A Reflexive View of the Expert-Lay Knowledge Divide." In *Risk, Environment and Modernity: Towards a New Ecology*, edited by Scott Lash, Bronislaw Szerszynski, and Brian Wynne, 44–83. London: Sage, 1996.

Wynne, Brian. "Misunderstood Misunderstanding: Social Identities and Public Uptake of Science." *Public Understanding of Science* 1, no. 3 (1992): 281–304.

Wynne, Brian. "Unruly Technology: Practical Rules, Impractical Discourses and Public Understanding." *Social Studies of Science* 18, no. 1 (1988): 147–67.

Index

Note: Page references in *italics* refer to figures.